Low Grade Heat Driv[illegible] Distillation and Desalination

Low Grade Heat Driven Multi-Effect Distillation and Desalination

Bijan Rahimi
School of Mechanical Engineering, Sharif University of Technology, Tehran, Iran

Hui Tong Chua
School of Mechanical and Chemical Engineering, The University of Western Australia, Perth, WA, Australia

ELSEVIER elsevier.com

Elsevier
Radarweg 29, PO Box 211, 1000 AE Amsterdam, Netherlands
The Boulevard, Langford Lane, Kidlington, Oxford OX5 1GB, United Kingdom
50 Hampshire Street, 5th Floor, Cambridge, MA 02139, United States

Library of Congress Cataloging-in-Publication Data
A catalog record for this book is available from the Library of Congress

British Library Cataloguing-in-Publication Data
A catalogue record for this book is available from the British Library

ISBN: 978-0-12-805124-5

Publisher: John Fedor
Acquisition Editor: Anita Koch
Editorial Project Manager: Sarah Jane Watson
Production Project Manager: Vijayaraj Purushothaman
Designer: Greg Harris

Reason says: The six directions are the boundary and there is no way out.
Love says: There is a way and I have many times traveled it.

Rumi, Persian poet and Sufi mystic, 13th century

Contents

"Excel spreadsheet package files" of this book is available in the following link: https://www.elsevier.com/books-and-journals/book-companion/9780128051245

Biography

Dr. Bijan Rahimi is a visiting assistant professor at the School of Mechanical Engineering of Sharif University of Technology (SUT). He is also a research leader of a novel desalination pilot plant project at the Institute of Water and Energy of SUT.

He recently received Iran's National Elites Foundation Award for his work on low grade heat-driven desalination and its application in Iran. He is also collaborating with the Water Research Institute of Iran's Ministry of Energy on the formulation of the first desalination research and technology roadmap for Iran.

He received his PhD in desalination from the School of Mechanical and Chemical Engineering of the University of Western Australia in 2016. He also collaborated with a major alumina refinery in Western Australia on the use of process waste heat to reconcentrate spent process liquor.

Besides academic experience, he has approximately 6 years of working experience in the field of waste heat management, renewable energy, measurement instrumentation, and control valves in the Iranian industries (oil and gas, chemical and petrochemical, steel, and paper sectors).

Bijan is a reviewer of the Desalination journal. He is a member of the Desalination Research Group at the Sharif University of Technology.

His ResearcherID (Web of Science) is D-6157-2013 and ORCID ID is 0000-0002-5788-8018.

For more information about his publications you can follow him here:

ResearchGate: https://www.researchgate.net/profile/Bijan_Rahimi.

LinkedIn: https://www.linkedin.com/in/bijan-rahimi-020a4a3b; https://ir.linkedin.com/in/bijan-rahimi-020a4a3b.

Professor Dr. Hui Tong Chua is a professor of Mechanical and Chemical Engineering and the Chemical Engineering Program Chair at the University of Western Australia (UWA). His research interest covers heat and mass transfer, thermodynamics, process engineering, and waste heat utilization. Seven of his international peer-reviewed journal articles are among the top 1% in the engineering field in terms of citation.

His research team invented the boosted and flash-boosted multi-effect distillation desalination processes in order to effectively couple with low grade waste heat. His team has also successfully demonstrated their principal ideas through a pilot plant. Through his ongoing collaboration with a major alumina refinery in Western Australia, his team has demonstrated the potential of the technologies toward significantly improving the energy and cost efficiencies of mineral refineries. His long-term vision is to realize this demonstrated potential.

Hui Tong is also a China Shanxi Province Hundred Man Distinguished Professor at the Taiyuan University of Technology, Shanxi Province, China.

Hui Tong holds a PhD, an M. Eng., and a B. Eng. (1st Class Hons.) in mechanical engineering from the National University of Singapore.

Prior to joining UWA, Hui Tong was an assistant professor at the Faculty of Engineering of the National University of Singapore.

Hui Tong's ResearcherID (Web of Science) is B-1317-2008.

For more information you can follow him here:

http://www.web.uwa.edu.au/people/huitong.chua.

ResearchGate: https://www.researchgate.net/profile/Hui_Chua.

Preface

This book is the result of 8 years of work, which started at a time when the Western Australian Geothermal Center of Excellence (WAGCoE) was first established and Hui Tong was its Aboveground Engineering program leader.

He has had a long history of working on the utilization of waste heat on heat-driven chillers, so that when he first encountered geothermal energy he quickly realized that in terms of its thermodynamic nature, it was a type of waste heat. As soon as the geothermal fluid is extracted from underground, we hope to utilize it to such an extent that ideally its temperature approaches the ambient surrounding's temperature before it is eventually reinjected underground.

Hui Tong and Prof. Klaus Regenauer-Lieb, currently the head of the School of Petroleum at the University of New South Wales, but then director of WAGCoE, worked very closely on the concept of geothermal desalination, given that Western Australia is water short and enjoys an abundance of underground aquifers.

Aided by the detailed catalog data of Alfa Laval's freshwater generators, Hui Tong came up with the idea of boosted multi-effect distillation (B-MED); this was subsequently consolidated by Klaus and Dr. Xiaolin Wang, currently a senior lecturer at the University of Tasmania. Hui Tong is gratified that the University of Western Australia has a strong patent position on this technology and holds a US Patent on this technology, principally through the fantastic support and council by Neil Prentice, Tom Schnepple, and Tymen Brom. Tymen is the Commercialization manager of the National Center of Excellence for Desalination Australia (NCEDA).

Funded by NCEDA and South32, Hui Tong's team, notably Dr. Alexander Christ, successfully developed a 1.5 m^3/day pilot plant to demonstrate the principal idea of the B-MED process. Dr. Bijan Rahimi, currently a visiting assistant professor at the Sharif University of Technology (SUT), joined the team in February 2012, and together with Hui Tong and Alex developed the flash-boosted multi-effect distillation (FB-MED) process.

Hui Tong's team worked very closely with Steve Rosenberg of ProChemistry Consulting, and Eric Boom and Silvio Nicoli of South32 for many years with the common goal of applying the distillation technologies to assist with the operation of alumina refineries. It was Silvio who realized that the FB-MED process could make a significant difference to the energy efficiency of alumina refineries and mineral refineries in general.

Hui Tong also wants to specially acknowledge the years of support offered by Paul Tuckwell of Alfa Laval. Significantly Paul has made available the picture of an Alfa Laval hot water–driven multi-effect distillation desalination plant as the cover image of this book.

Hui Tong and Bijan would like to acknowledge all those named for their friendship and support, without which this book will certainly be impossible. Obviously Hui Tong and Bijan take full responsibility for any mistakes and omissions in this book.

Bijan is currently spearheading the prototyping of the FB-MED process at SUT. He wants to thank Prof. Ali. A. Alamolhoda, Prof. Majid Abbaspour, and Prof. Hamid Mehdigholi for assisting him in making the prototyping effort possible.

Most importantly Hui Tong wants to express his gratitude and respect to Prof. Jeffrey M. Gordon for all the years of guidance, supervision, mentorship, and constant encouragement. Jeff is a fatherly figure to Hui Tong.

Bijan Rahimi (bijan@mehr.sharif.ir)
Sharif University of Technology

Hui Tong Chua (huitong.chua@uwa.edu.au)
The University of Western Australia

CHAPTER 1

Introduction to Desalination

1.1 Introduction

Nearly 71% of the surface of the earth (510×10^6 km^2) is covered by the oceans and the remaining 29% is covered by land [1]. There is certainly bountiful water available on earth, but only 3% is drinkable and 97% is saltwater [2]. Nearly 70% of this available freshwater is frozen in glaciers, while the remaining 30% is in underground hard-to-reach aquifers, of which approximately 0.25% flows into rivers and lakes for direct use [3]. Therefore, traditional sources of available freshwater such as underground aquifers and surface water constitute a limited quantity worldwide. Furthermore, depletion of these sources is increasing at an alarming rate [4].

Water scarcity is the mismatch of demand and availability of freshwater in a particular location. It has become a worldwide issue with the pollution of existing water supplies, increasing population and industry activity, uneven freshwater to population distributions, and changing rainfall patterns. This implies that many regions containing populated centers are becoming less capable of meeting the water supply requirements of the residing populations [3,5–7]. Water-stressed countries currently encompass one-third of the world's population and it is predicted to reach two-thirds by 2025 [8]. Aside from residential and industrial water shortage, agriculture is also being affected directly by water shortage. Farmers increasingly have to compete for water with urban residents and industries, thereby placing global food security at risk [9].

Methods of attenuating such water supply issues include wastewater treatment and reuse, desalination, as well as water conservation schemes. Some 80 countries face severe water shortage [10], while some countries such as Kuwait, the United Arab Emirates, and Saudi Arabia currently depend almost entirely on desalination for their supply of water [4].

As a result of these situations, seawater desalination has become an essential option to augment freshwater resources, especially in developing countries and many arid zones. As an example, in 2010 the Gulf Cooperation Council (GCC) countries (in the Middle East region) produce around 39% of the world's desalinated water production [11,12].

The desalination process is being increasingly adopted over traditional water supply methods because the cost per unit volume of water produced has come down for desalination while it has risen for traditional methods [4,7,13]. In 2011, approximately 150

Low Grade Heat Driven Multi-Effect Distillation and Desalination. http://dx.doi.org/10.1016/B978-0-12-805124-5.00001-2

countries worldwide used around 15,988 desalination plants (these include online, under construction, and contracted) to produce desalinated water [14]. The total global capacity of all online plants was 70.8 Mm^3/day in 2011 [15]. This is a 10% increase compared to capacity in 2010. Also, 632 new plants were added from mid-2011 to August of 2012, thereby increasing the installed capacity to 74.8 Mm^3/day [15]. As of June 30, 2015 the total number of desalination plants worldwide reached 18,426 with a total production rate exceeding 86.8 Mm^3/day, which satisfied the need of around 300 million people around the world [16]. These data indicate the potential of the desalination market in both aspects of freshwater production rate and energy consumption.

1.2 A Brief History of Desalination

The word origin and history of desalination dates back to 1943, and as a verb, "desalt" was recorded in 1909 [17]. The concept of desalination is much older, however, with references to it being found in ancient writings [18]. Historically, salt has been held as a precious commodity. The first goal of desalination was not related to producing freshwater, but rather to extract and use the salt from salty water by means of natural evaporation [2].

It is difficult to pinpoint the first instance when humans desalinated salty water for freshwater, but Aristotle (384−322 BC) was one of the earliest recorded scientists who explained the desalination process. His understanding was based on his observations that when saltwater turns into vapor, the condensed vapor does not contain any salt [19]. At that time, the needs for producing freshwater for sailors was critical for long distance voyages. Ancient drawings depict sailors boiling seawater and suspending a large sponge from the mouth of a brass vessel to absorb what is evaporated [3]. Therefore, it is reasonable that the interest in desalination dates back as far as the 4th century BC.

Advanced technologies that mimic natural processes such as evaporation-condensation or osmosis to obtain freshwater from seawater, have been developed only in the modern decades. Basic desalination processes were used on naval ships from the 17th to 19th centuries. For example, in 1790, the US Secretary of State, Thomas Jefferson, received an offer to sell the government a seawater desalination scheme [2]. The first desalination units were eventually built for ships that were used to provide fresh boiler water, thereby removing the need to travel with cargo loads of water [18]. Years later, a British patent was granted in 1852 [20], and then in 1872 the first solar still was designed by a Swedish engineer, Carlos Wilson, and constructed in Chile [21]. In 1912, a 75 m^3/day desalination plant was installed in Egypt [22]. The island of Curacao in the Netherlands Antilles was the first location to make a major commitment to desalination in 1928, followed by a major seawater desalination plant built in Saudi Arabia in 1938 [2,23]. During that period (1929–37), the total desalinated water capacity escalated due to the emergence of the oil industry [22].

In the 1940s, during World War II, research on desalination was conducted to find the proper ways to meet military requirements for freshwater in regions where soldiers were facing drinking water shortage [2]. For example, Telkes [24] developed a plastic still inflated with air for desalination, which was used by the US Air Force and US Navy during World War II. After World War II, the United States and other countries continued their work on desalination. The US Congress passed the Saline Water Conversion Act (PL 82-448) in 1952, which created and funded the Office of Saline Water within the Department of the Interior's Bureau of Reclamation [2].

In the 1960s, desalination science entered into a new and modern era; it was a special time for commercialization because of the dramatic growth of population and water shortages experienced worldwide. New methods of desalination were considered based on fossil resources, because many oil-rich countries in the Middle East and North Africa (MENA) region had been faced with water shortages and therefore preferred to dedicate part of their natural energy resources (oil and gas) to their local water desalination production instead of exportation [11]. Recently, desalinated water has become a commodity for many countries and desalination plants are not limited to the MENA region alone [25].

The first generation of desalination plants was commissioned in Shuwaikh, Kuwait and in Guernsey, Channel Island in 1960 [22]. By the late 1960s, desalination plants that could produce up to 8000 m^3/day were beginning to be installed in various parts of the world and were mostly based on thermal process, which was expensive as it required a lot of energy [26]. Nevertheless, they were good enough for the Middle Eastern oil-rich countries. Since the 1970s, membrane processes have been used extensively and commercialized for large-scale production [22,26]. Finally, in the 1980s, desalination became a fully commercial enterprise and this continues today [26]. According to the International Desalination Association report [15], around 18,000 desalination plants are in operation worldwide with an approximate capacity of 90 Mm^3/day of freshwater. The main worldwide use of desalinated water is for municipal and industrial purposes.

1.3 Desalination Technologies

In general all applicable desalination processes can be divided into two main categories based on the phase change of saline feed water.

- **Desalination with phase change**: This category includes all heat-driven processes where freshwater is produced by evaporation and condensation phenomena.

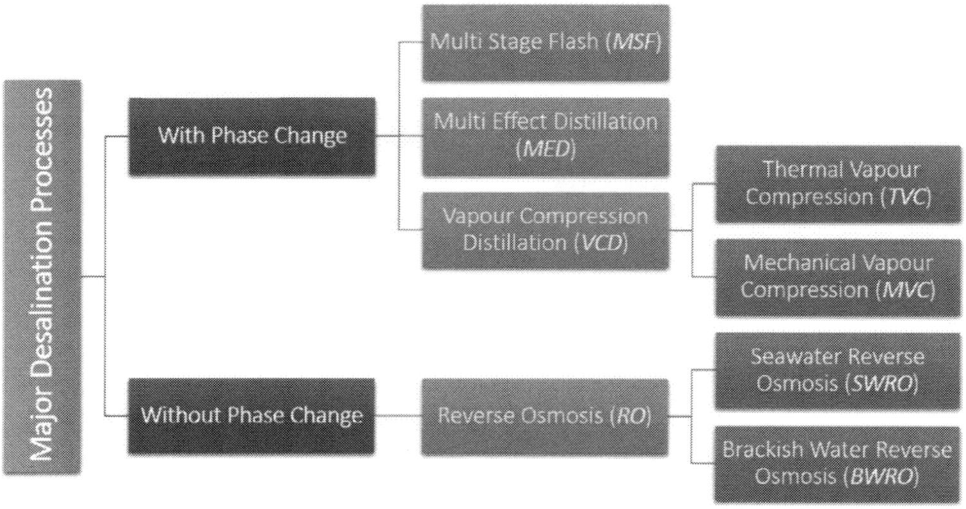

Figure 1.1
Major desalination processes.

- **Desalination without phase change**: In this category, separation is achieved by passing saline water through membranes without involving phase change; reverse osmosis (RO) is the iconic example.

Fig. 1.1 shows the most dominant methods for seawater desalination. Selecting the proper method for saline water desalination depends on many factors, such as the location of the plant, plant capacity, plant life, initial costs, interest rate, required freshwater quality, type of energy sources, plant load factor, and water cost. In industrial applications, for which the main aim is to extract freshwater from the process of liquid streams, it should be determined whether the selected process can handle the liquid or not. For instance, in mining industries and refineries such as alumina refineries, thermal desalination processes such as multi-effect distillation (MED) and multi-stage flash (MSF) are more suitable than RO processes, which cannot handle the process streams because of significant fouling. As an example, in the evaporation units of an alumina refinery plant (see Chapter 9), RO technologies cannot extract the freshwater from the process liquor (which can be considered as equivalent to 20 wt% caustic soda solution); therefore the only option is to use thermal technologies.

1.3.1 Processes With Phase Change

The principal examples for processes with phase separation are MSF, MED or multi-effect evaporation (MEE), and vapor compression distillation (VCD), including thermal vapor compression (TVC) and mechanical vapor compression (MVC), which will be explained in Sections 1.3.1.1–1.3.1.3 [4,22,26,27].

These methods mimic the natural method of desalination using evaporation-condensation phenomena for producing freshwater from saline waters (or industrial process liquid streams). Due to the greater availability and lower cost of fossil fuels in the Middle East, the use of thermal desalination processes such as MSF is concentrated primarily within these regions [28]. In 1996, 10,000 tons/year of oil was required to produce 1000 m^3/day of freshwater [29]. In August 2012, the share of MSF and MED technologies of the total installed capacity for saline water desalination applications was around 31% [15]. These two technologies are the major technologies used in GCC countries, including 68% of the desalination capacities in 2012 [11,30]. The other reason thermal phase change processes are more applicable in that region is that Persian Gulf water is known as 4H [11]: high salinity (around 45,000 ppm [2]), high turbidity, high temperature, and high marine life. However, some GCC countries such as Oman and the Kingdom of Saudi Arabia use RO for desalting the Indian Ocean and Red Sea waters, which are more suitable feeds for RO than the Persian Gulf water [11].

1.3.1.1 Multi-Stage Flash

The MSF distillation process (Fig. 1.2) accounts for the major portion of desalinated municipal drinking water produced in the MENA region [22]. Today it stands in second place in the world after RO technology and is used primarily for desalting seawater [15]. This process has been in large-scale commercial use for over 30 years, and because of a higher resistance against scaling as compare to MED technology, it has continually displaced the MED process from the 1950s (since its development time) [25]. In 1957, four MSF plants with the total capacity of 9084 m^3/day were installed in Kuwait [31]. The Al-Jubail plant (815,120 m^3/day) in Saudi Arabia is the largest MSF plant in the world [32].

In this technology, as shown in Fig. 1.2, seawater feeds are pressurized, heated, and discharged into a series of flashing chambers maintained slightly below the saturation vapor pressure, therefore a fraction of the feed water flashes into steam. The flashed steam passes through a mist eliminator and condenses on the exterior surface of the heat transfer

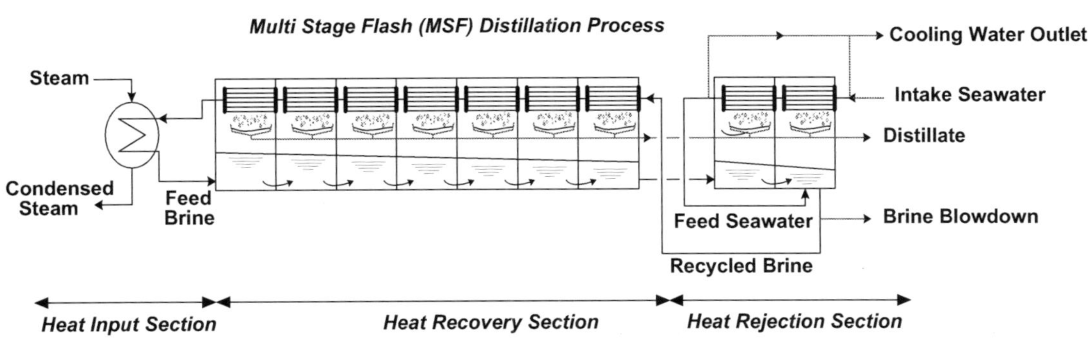

Figure 1.2
A standard (conventional) multi-stage flash schematic design.

tubing (condenser) that is located at the top of each stage. The condensed liquid then drips into trays and is collected as freshwater. All MSF plants consist of a heat source, heat recovery, and heat rejection sections [33]; and typically their heat recovery sections consist of 19–28 stages (in modern and large MSF plants) [4]. The heat rejection section usually consists of three or more stages, which are used to control the recycled brine temperature [33] and to conserve antiscaling agents. Top brine temperature (TBT) in the heat input section is typically between 90°C and 110°C [33,34]; however, it can reach to between 120°C and 130°C with some modifications [35,36]. Fig. 1.2 shows the conventional MSF schematic design. Typical specific pumping power consumption for this technology is mostly between 3.0 and 5.0 kWh/m^3 for large modern MSF plants [11,37]. The main advantages and disadvantages of this technology are listed here [4,22,25–28,38].

Advantages:

- Produce large capacities of freshwater
- Independent of the salinity of the feed water
- Easy operation and very low performance degradation within the plant life
- Very high quality freshwater (less than 10 mg/L TDS; total dissolved solids)
- Minimal requirement for pretreatment of the feed water compared to the RO process
- Simple operational and maintenance procedures
- Higher resistance against scaling compared to the MED process
- Long background of commercial use and reliability
- The possibility for being combined with other processes

Disadvantages:

- Expensive to build and operate
- Requires a high level of technical knowledge
- Highly thermal energy intensive process
- Operates at high temperature (top brine temperatures of 90–120°C)
- Low recovery ratio, means more feed water is required to produce the same amount of freshwater compared to the MED process
- Turndown ratio limited to about 70%

1.3.1.2 Multi-Effect Distillation or Evaporation

MED or MEE technology is the oldest industrial desalination method [39] and the most efficient thermal distillation process thermodynamically [40], although it stands in second place after MSF in the thermal desalination market [15]. It is called MED when the distillate is the main product of this process, and called MEE if evaporation is the main goal (e.g., to concentrate the feed stream); however the principle of both processes is the same. In Chapter 8, this process is termed MED because the process is used for seawater

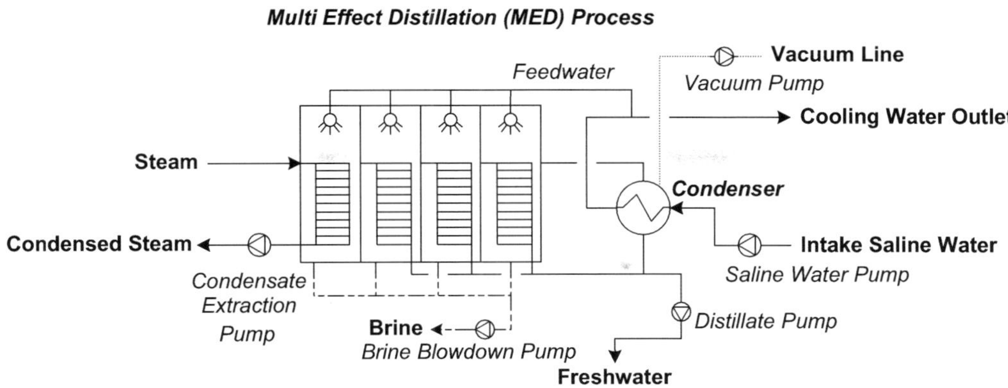

Figure 1.3
A standard (conventional) multi-effect distillation schematic design for seawater application with horizontal falling film evaporator.

desalination application. In contrast, in Chapter 9, when the main goal is to remove water from a mineral refinery process stream, it is termed the MEE process.

The capacity of MED plants is typically more than 300 m^3/day since anything less is not financially viable [27]. In this system (Fig. 1.3) the feed water is distributed onto the heat exchanger surfaces of the first effect. The heat source fluid (steam or hot liquid) that flows through the heat exchanger releases its energy to the distributed feed water and evaporates a portion of the feed water. The produced vapor then condenses in the heat exchanger of the second effect to evaporate more feed water in that effect. The brine from the first effect is then purged. At the second effect, the evaporated feed water goes on to power the third effect with the resulting brine being drained from the bottom of that effect. This process continues to the last effect with the corresponding produced vapor entering the condenser section and condensed by the incoming saline feed water acting as a coolant. Part of the preheated saline water is then sent to the various effects as feed water [41].

MEE processes have many possible configurations. For example, with respect to the evaporator design, it can use a rising film vertical or a falling film horizontal tube. With respect to the feed flow, it can be backward, parallel, or forward. The effects arrangement, in turn, can be horizontal or use a stacked layout [39].

The main difference between MED and MSF is related to the difference between boiling and flashing evaporation phenomena. Flashing technique (MSF process) requires much more feed water and pressure difference compared to the boiling technique (MED process) for producing the same amount of vapor. This means that the flashing technique consumes much more pumping power. Typical specific pumping power consumption for large-scale

MED plants is around 1.5 to 3 kWh/m^3 [4,11], which is significantly lower than the MSF process.

The other difference is related to the top brine temperature (TBT) and scaling issue. Because of the design of MED processes, some cleaning procedures that can be applied to MSF cannot be used in MED, thus, the best way to manage scaling is to keep the TBT to around 65−70°C [42].

The main advantages and disadvantages of this technology are listed here [22,27,33,38,39].

Advantages:

- Operates at low temperature
- High quality freshwater
- High thermal performance
- The possibility to be combined with other processes
- Low pumping power
- Very reliable process
- Handle normal levels of biological or suspended matter
- Minimum pretreatment requirements
- Minimum labor requirement
- Enjoys a full turndown ratio

Disadvantages:

- Expensive to build (high capital cost)
- Susceptible to corrosion
- Low recovery ratio (though not as low as MSF)

1.3.1.3 Vapor Compression Distillation

The VCD process is mostly used for small- to medium-scale desalination plants [43]. They are often used in industries where freshwater is not readily available [26]. Generally, the heat for evaporation in VCD processes is provided by compression. TVC and MVC are the two configurations of VCD technology. In TVC processes, the vapor is compressed by a thermo-compressor, which is typically driven by medium-pressure live steam. In contrast, in MVC processes, the vapor is compressed by a mechanically driven compressor that uses only electricity to operate; therefore, in cases where steam is not available as a heat source, this process can be an option.

VCD technology also can be used in combination with MED and MSF processes, such as TVC-MED [44] or TVC-MSF [34]; a combination that can increase the system performance ratio over the standalone MED or MSF system [22,34]. TVC-MED and MVC-MED

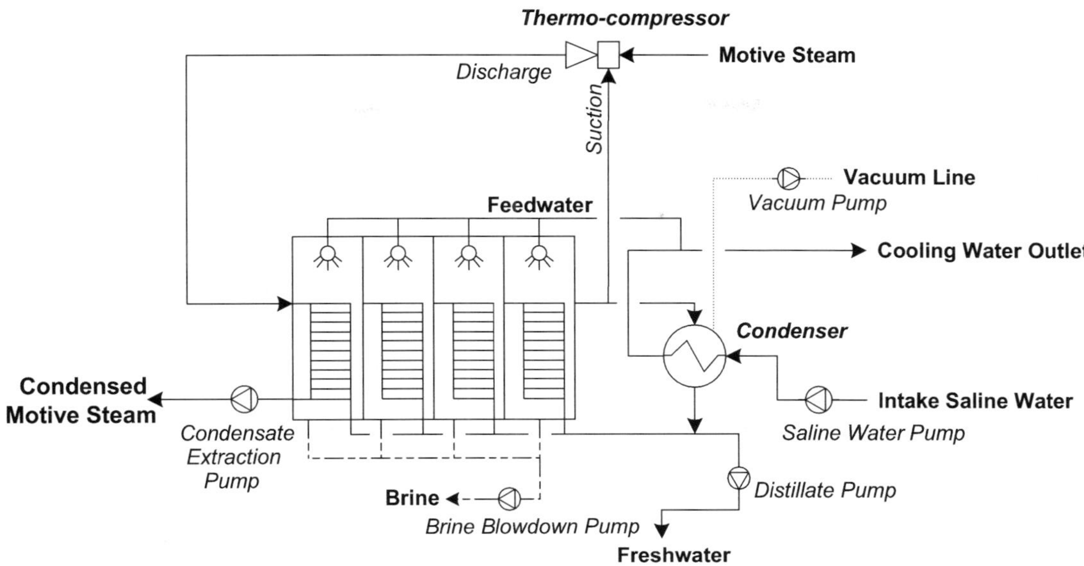

Figure 1.4
Schematic design of a thermal vapor compression multi-effect distillation process for seawater application with horizontal falling film evaporators.

(Figs. 1.4 and 1.5) are the most dominant processes due to the higher thermal performance of MED as compared with MSF processes with the TVC-MED system enjoying the highest thermal performance among all steam-driven processes. As shown in Fig. 1.5, in MVC-MED processes the produced vapor in the last effect is compressed in a mechanical compressor so that this system is attractive in places where cooling sources are not available.

Typically an MVC-MED plant's capacity range is up to 3000 m^3/day [4,43], and the range for a TVC-MED plant is up to about 20,000 m^3/day [43]. In order to reach higher capacities, typically two or more plants will have to work in parallel. For example, Curaçao Island 12,000 m^3/day (1994), Umm Al Nar (UAE) 16,000 m^3/day (1998), Layyah (UAE) 36,370 m^3/day (2005), and Fujairah II (UAE) 38,670 m^3/day (2007) are utilizing TVC-MED [45].

The electrical power consumption of MVC-MED units is higher than that of TVC-MED (as mentioned earlier) and is typically between 7 and 12 kWh/m^3 [26]. TVC-MED's electrical consumption is typically around 1 kWh/m^3 [39], which is even lower than that of conventional MED. In contrast, MVC-MED does not require heat sources.

As mentioned earlier, TVC-MED and MVC-MED are the most dominant steam-driven processes as compared to other thermal desalination processes such as MSF and MED. Their main advantages and disadvantages are listed here [22,46].

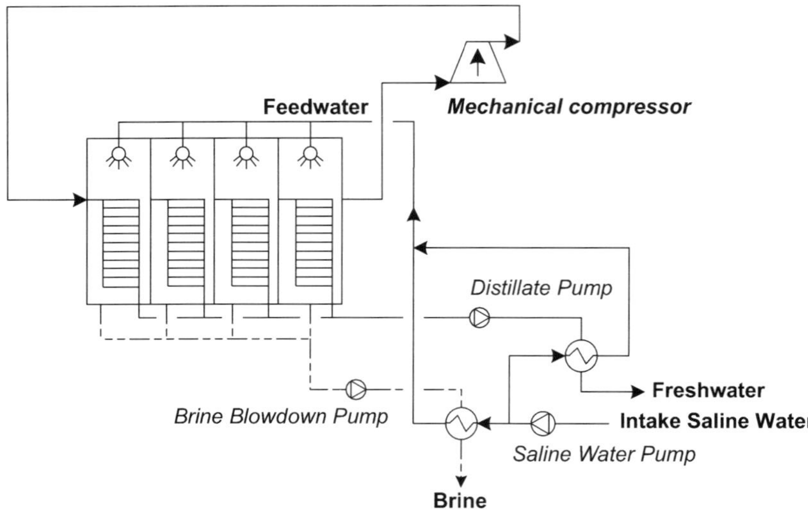

Figure 1.5
Schematic design of a mechanical vapor compression multi-effect distillation process for seawater application with horizontal falling film evaporators.

Advantages:

- Operates at low temperature
- High quality freshwater
- No need for steam line and cooling source (for MVC-MED)
- Highest thermal performance (for TVC-MED)
- Lowest pumping power (for TVC-MED)
- Very reliable process
- Thermocompressor is very robust (for TVC-MED)
- Minimum labor requirement

Disadvantages:

- Expensive to build (high capital costs)
- Susceptible to corrosion
- High pumping power (for MVC-MED)
- Higher investment cost due to the cost of mechanical compressor (for MVC-MED as compared to MED and TVC-MED)

1.3.2 Processes Without Phase Change

In these processes, freshwater is extracted by passing saline water through membranes without involving phase change; RO is the main technology in this category. Membrane-based desalination plants, which make up the majority of plants worldwide [13], are largely powered by electricity derived from nonrenewable sources.

An osmosis phenomenon is the transfer of the solvent (pure water) of a solution (saline water) based on a concentration gradient, through a semipermeable membrane. For example, if freshwater and saline water are separated by a semipermeable membrane, due to the concentration gradient between the two medium, freshwater (low concentrate medium) will move toward the saline water (high concentrate medium) to achieve an overall concentration balance (equilibrium) in the system. The principle behind this movement is osmotic pressure. Therefore by gradually increasing the pressure on the saline water side (against the osmotic pressure direction), until a point whereby the applied pressure equals the osmotic pressure, the water flow across the membrane will be stopped. At this point, any further increase in the pressure on the side of the saline water will drive the flow in the opposite direction so that instead of moving freshwater toward the saline water, the freshwater can be separated from the saline water. This process is called reverse osmosis. The pressure difference between the applied pressure and the osmotic pressure is one of the key factors in the RO process to identify the freshwater mass flow rate that passes through the membrane (Fig. 1.6) [47]. Depending on the feed water quality, the technology, and the type of membranes, around 30–80% of the total feed water can be desalinated [27].

Typically RO processes include four major sections, namely pretreatment, high-pressure pumps, membranes, and post-treatment (Fig. 1.7). A single-stage RO-produced water has a salinity of less than 500 mg/L TDS [27]. Pretreatment of the feed water is an essential component of the RO plant so as to prevent scaling of the membranes. Because there is no

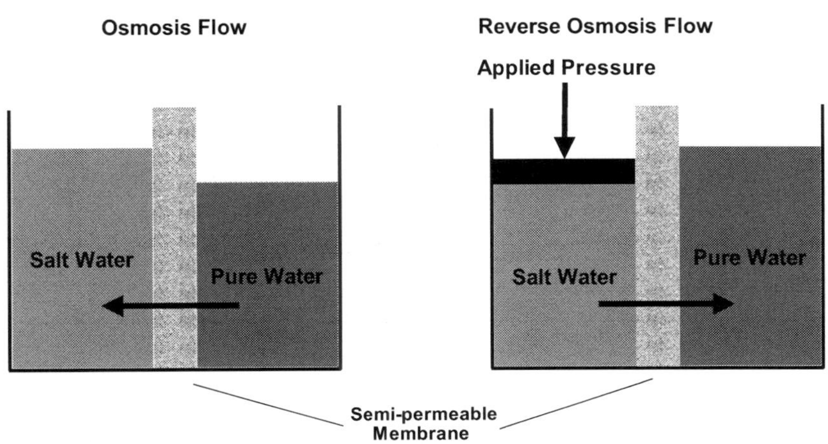

Figure 1.6
Osmosis and reverse osmosis flows.

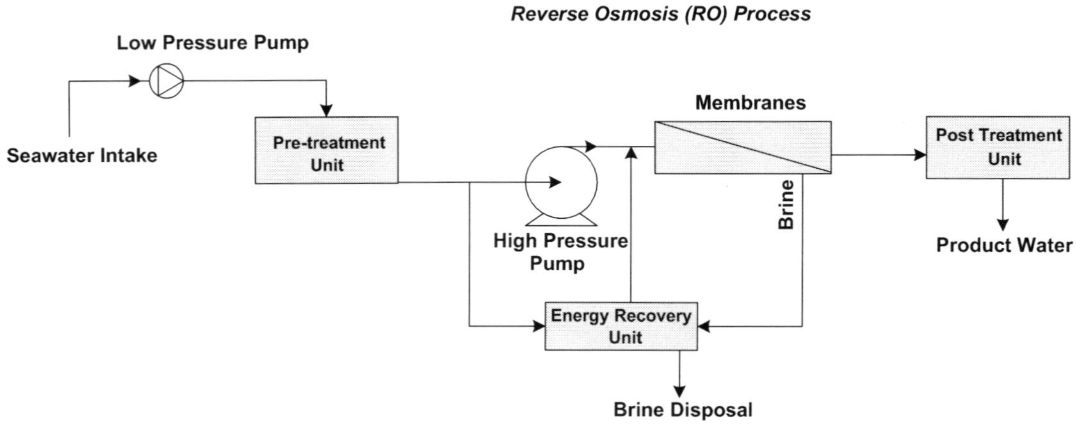

Figure 1.7
Schematic design of a typical reverse osmosis (RO) plant.

requirement to effect phase change, the major use of energy for RO is to pressurize the feed water. The typical specific electrical consumption of RO plants is between 3 and 7 kWh/m^3 for seawater application [48–50].

RO is the fastest growing desalination process compared to other desalination technologies due to the reduction in membrane costs and developments of energy recovery devices [51]. Its share of installed capacities was 63% in 2012 [15].

The main advantages and disadvantages of this technology are listed here [22,27,38].

Advantages:

- Quick to build and simple to operate
- Easy to extend the capacity by adding on extra modules
- Low energy consumption
- No need to heat sources
- Lower usage of chemicals for cleaning purposes
- No need to shut down the entire plant for scheduled maintenance

Disadvantages:

- Expensive membranes and low life expectancy (2–5 years)
- Sensitive toward changes in feed water salinity
- Need for high quality standard of materials and equipment
- Bacterial contamination is possible
- Pretreatment of the feed water is essential
- High-pressure operation, which causes mechanical failure

1.4 Energy Consumption and Environmental Impacts of Desalination Processes

Desalination is considered an energy-intensive process, however it represents only a small fraction of the total national energy consumption [50]. As a very general figure, energy consumption for desalinated water in a household environment is only around 3.2% of the total family energy consumption, including electricity, gas, and fuels [52].

Desalination processes predominantly require large amounts of electrical and thermal energy. Concurrently water desalination, by both thermal and membrane processes, is becoming cheaper owing to material improvements, process improvements, and an increase in competition [6].

The global rate of installation of water desalination facilities is increasing, with estimates of an approximate 55% increase in the number of plants installed in 2012 over the previous year [6]. The increased adoption of desalination represents a significant rise in global energy usage, due to the energy-intensive processes required. For a perspective, consider that the total amount of energy used for desalination worldwide has become comparable to the total energy requirement of a small industrialized country such as Sweden [53].

Around 30–50% of the product water cost is related to the energy cost of the desalination process [50]; this share makes the energy saving issue an important key factor for water price reduction. Therefore, any improvements into the energy utilization and efficiency of desalination processes are important, particularly those that incorporate the use of industrial waste heat or renewable energy sources such as geothermal energy, and solar energy [54].

The minimum theoretical energy for desalination is the minimum required isothermal reversible work of separation to produce freshwater from seawater, which is independent of the process and is a function of the concentration and recovery factor. The theoretical threshold for 35,000 ppm NaCl solution at zero recovery is equal to 0.79 kWh/m^3 [52]; the value for a typical recovery of 50% is changed to 1.06 kWh/m^3 [55]. The actual energy consumption for the desalination process is larger than the quantities due to all the irreversibilities related to the nature of the utilized desalination process. For example in thermal desalination plants the majority of these irreversibilities are related to heat transfer. MSF and MED processes consume both thermal and electrical energy. For typical MSF plants with the maximum live steam (as heat source) temperature of 120°C, the thermal energy consumption is around 12 kWh/m^3 of produced freshwater, while it is around 6 kWh/m^3 for an MED plant, which operates at lower temperatures (less than 70°C) [56]. The electrical consumption of these processes also should be considered to find the overall energy consumption. In contrast, RO processes only consume electrical energy in a range between 3 and 7 kWh/m^3 for seawater application [50,52,56,57]. Thus, the overall energy consumption of RO processes is closer to the minimum theoretical consumption threshold compared to MED and MSF processes. However, it does not mean that the RO's unit product cost is necessarily cheaper than the thermal processes. In addition to energy consumption, there are many other issues that can influence the product costs. These will be discussed further in Chapter 7.

As mentioned before, 30–50% of the water cost is related to the energy consumption of desalination processes, which is related to both electrical and thermal energy. For example, the RO process has a high overall efficiency at the expense of consuming a large amount of electricity, especially for running the high pressure pumps. For thermal processes, such as MSF and MED, it is related to large thermal energy consumption, besides the electrical energy consumption. Therefore with respect to production rate, economic feasibility, and environmental friendliness, the optimization of desalination methods should be considered in the context of minimizing energy consumption [41]. In this regard, incorporating renewable energy sources into desalination has been shown to be both technically and economically feasible and it should be considered an incontestable goal of ongoing research into desalination technology to incorporate renewable energy technologies henceforth [54].

Though more economical than renewable energy sources, the use of fossil fuels should also be framed within the context of environmental implications including the greenhouse effect, carbon dioxide emissions, and associated environmental pollution. Both energy and environmental issues are linked together because the energy generation processes that produce the required energy for desalination processes has potential environmental impacts. However, the design and management of the desalination process by itself can also have environmental impacts [50]. For thermal processes such as MSF and MED, discharging fumes such as CO_2, NO_x, and SO_x, and the rejected concentrated brine, are the two major impact factors on the environment [58]. Fumes are directly connected to the energy supply processes, so any use of alternative (sustainable) energy sources such as industrial waste heat, geothermal energy, or solar energy instead of fossil fuels can substantially reduce these greenhouse gas emissions. As a perspective, around 1.4–1.8 kg CO_2 is emitted per cubic meter of freshwater in the current state-of-the-art seawater RO plants [55,59–61].

The environmental impacts of desalination plants are not limited to greenhouse gasses. Intake seawater systems, brine disposal processes, and energy dissipated (thermal impact) in the sea by thermal processes can affect the marine life and give rise to local elevation of salinity and temperature [56]. An intake seawater system is associated with the impingement and entrainment of marine organisms [2,23,56]. By locating low-velocity open surface intakes with a combination of appropriate meshed screens into deeper offshore waters and reducing the volume of intake water by reusing the power plant's cooling water as feed for the desalination plant, the impacts from impingement and entrainment problems could be substantially minimized [23,55,56]. Brine rejection into the ocean can increase the local temperature and salinity. Many studies have shown that elevated salinity has little or no environmental impact [50] and prediluting the rejected brine with other waste streams such as power plant cooling water can solve the high salinity impact problem [56]. Pretreatment, post-treatment, and washing processes reject some chemicals such as heavy metals, antifoams, antiscalant, coagulants, and cleaning chemicals to the ocean, and tracing the impact of these chemicals into the environment is an important issue that is recognized as an environmental hazard [62]. To avoid the impacts of thermal rejection, the maximum heat dissipation is conducted from the waste streams before entering the sea [56]. Sommariva et al. [58] demonstrated a clear relationship between the plant efficiency and thermal environmental impact—the higher the plant efficiency, the lower the environmental impact.

Much research has been conducted on the environmental impact of desalination [50,52,55–58,63], and ISO 14000 is an assessment tool for the environmental behavior of desalination plants, but these are still in their early development phase [58].

References

[1] I.A. Shiklomanov, J.C. Rodda, World Water Resources at the Beginning of the Twenty-First Century, in: International Hydrology Series 2003, Cambridge University Press, 2003.

[2] H. Cooley, P.H. Gleick, G. Wolff, Desalination, with a Grain of Salt a California Perspective, Pacific Institute for Studies in Development, Environment, and Security, Oakland, California, June 2006.

[3] S. Kalogirou, Seawater desalination using renewable energy sources, Prog. Energy Combust. Sci. 31 (3) (2005) 242–281.

[4] A.D. Khawaji, I.K. Kutubkhanah, J.-M. Wie, Advances in seawater desalination technologies, Desalination 221 (1–3) (March 2008) 47–69.

[5] T.F. Stocker, D. Qin, G.-K. Plattner, M.M.B. Tignor, S.K. Allen, J. Boschung, A. Nauels, Y. Xia, V. Bex, P.M. Midgley, Climate Change 2013 the Physical Science Basis Working Group I Contribution to the Fifth Assessment Report of the Intergovernmental Panel on Climate Change, Cambridge University Press, New York, USA, 2013.

[6] N. Ghaffour, T.M. Missimer, G.L. Amy, Technical review and evaluation of the economics of water desalination : current and future challenges for better water supply sustainability, Desalination 309 (2013) 197–207.

[7] R. Einav, K. Harussi, D. Periy, The footprint of the desalination processes on the environment, Desalination 152 (2002) 141–154.

[8] R. F. Service, Desalination freshens up, Science 313 (80) (2006) 1088–1090.

[9] M.W. Rosegrant, X. Cai, S.A. Cline, Global water outlook to 2025 averting an impending crisis, New Engl. J. Public Policy 21 (2) (2007) 102–127.

[10] I.C. Karagiannis, P.G. Soldatos, Water desalination cost literature: review and assessment, Desalination 223 (2008) 448–456.

[11] H. Fath, A. Sadik, T. Mezher, Present and future trend in the production and energy consumption of desalinated water in GCC countries, Int. J. Therm. Environ. Eng. 5 (2) (2013) 155–165.

[12] J.R. Hillamana, E. Baydoun, Overview of the roles of energy and water in addressing global food security, Int. J. Therm. Environ. Eng. 4 (2) (2012) 149–156.
[13] M.K. Wittholz, B.K. O'Neill, C.B. Colby, D. Lewis, Estimating the cost of desalination plants using a cost database, Desalination 229 (1–3) (September 2008) 10–20.
[14] L. Henthorne, T. Pankratz, S. Murphy, The state of desalination 2011, in: IDA World Congress on Desalination and Water Reuse, Desalination: Sustainable Solutions for a Thirsty Planet, 2011.
[15] T. Pankratz, IDA Desalination Yearbook 2012–2013, Media Analytics Ltd., Oxford, 2013.
[16] IDA, Desalination by the Numbers, 2016 [Online]. Available: http://idadesal.org/desalination-101/desalination-by-the-numbers/.
[17] D. Harper, Online Etymology Dictionary, 2015 [Online]. Available: http://www.etymonline.com/.
[18] Waterworks Museum, Back to basics with desalination, World Pumps 2012 (9) (2010) 32–34.
[19] E. Delyannis, Historic background of desalination and renewable energies, Sol. Energy 75 (5) (November 2003) 357–366.
[20] P. Simon, Tapped Out: The Coming World Crisis in Water and What We Can Do about It, Welcome Rain Publishers, New York, 1998.
[21] S. Parekh, M.M. Farid, J.R. Selman, S. Al-Hallaj, Solar desalination with a humidification-dehumidification technique - a comprehensive technical review, Desalination 160 (2004) 167–186.
[22] H.T. El-Dessouky, H.M. Ettouney, Fundamentals of Salt Water Desalination, Elsevier Science B.V., 2002.
[23] NRC, Desalination: A National Perspective. Committee on Advancing Desalination Technology, Water Science and Technology Board, National Research Council (NRC) of the U.S. National Academies, Washington D.C., 2008.
[24] M. Telkes, Solar Distiller for Life Rafts, US Office of Science, R&D Report No. 5225, P/B.21120, 1945.
[25] S. Lattemann, M.D. Kennedy, J.C. Schippers, G. Amy, Chapter 2 Global desalination situation, in: Sustainability Science and Engineering, Sustainable Water for the Future: Water Recycling Versus Desalination, vol. 2, Elsevier B.V., 2010, pp. 7–39.
[26] O.K. Buros, The ABCs of Desalting, second ed., USA: International Desalination Association (IDA), Massachusetts, 1999.
[27] AFFA, Introduction to Desalination Technologies in Australia, Department of Agriculture, Fisheries & Forestry – Australia, Turner, ACT, 2002.
[28] R. Borsani, S. Rebagliati, Fundamentals and costing of MSF desalination plants and comparison with other technologies, Desalination 182 (1–3) (November 2005) 29–37.
[29] S.A. Kalogirou, Solar energy for seawater desalination selection of the best system based on techno-economic factors, in: Mediterranean Conference on Renewable Energy Sources for Water Production, 1996, pp. 10–12.
[30] IDA, Desalting Plants Inventory, 2011 [Online]. Available: http://www.desaldata.com.
[31] K.S. Spiegler, A.D.K. Laird, Principles of Desalination, second ed., Academic Press, New York, 1980.
[32] A.M. Al-Mudaiheem, H. Miyamura, Construction and commissioning of Al Jobail phase II desalination plant, in: Second IDA World Congress on Desalination and Water Re-use, 1998, pp. 1–11.
[33] H. El-Dessouky, I. Alatiqi, H. Ettouney, Process synthesis: the multi-stage flash desalination system, Desalination 115 (2) (July 1998) 155–179.
[34] H. El-Dessouky, H. Ettouney, H. Al-Fulaij, F. Mandani, Multistage flash desalination combined with thermal vapor compression, Chem. Eng. Process. 39 (4) (July 2000) 343–356.
[35] A.M. Helal, M. Odeh, The once-through MSF design. Feasibility for future large capacity desalination plants, Desalination 166 (August 2004) 25–39.
[36] A.M. Helal, Once-through and brine recirculation MSF designs a comparative study, Desalination 171 (2004) 33–60.
[37] C. Sommariva, R. Borsani, M.I. Butt, A.H. Sultan, Reduction of power requirements for MSF desalination plants: the example of Al Taweelah B, Desalination 108 (1996) 37–42.
[38] ASIRC, KPMG, ATSE, Overview of Treatment Processes for the Production of Fit for Purpose Water: Desalination and Membrane Technologies, Australian Sustainable Industry Research Centre LTD, ASIRC Report No.: R05-2207, Monash University, Churchill, VIC, 2005, p. 3824.

[39] M. Al-Shammiri, M. Safar, Multi-effect distillation plants: state of the art, Desalination 126 (1−3) (November 1999) 45−59.
[40] A. Ophir, F. Lokiec, Advanced MED process for most economical sea water desalination, Desalination 182 (1−3) (November 2005) 187−198.
[41] B. Rahimi, A. Christ, K. Regenauer-Lieb, H.T. Chua, A novel process for low grade heat driven desalination, Desalination 351 (October 2014) 202−212.
[42] J. Laborie, Scaling in the Sea Water Evaporators, SIDEM Co., Veolia Environnement, France, 2004.
[43] A.A. Al-Karaghouli, L.L. Kazmerski, Renewable energy opportunities in water desalination, in: Desalination, Trends and Technologies, InTech, 2011, pp. 149−184.
[44] I.S. Al-Mutaz, I. Wazeer, Development of a steady-state mathematical model for MEE-TVC desalination plants, Desalination 351 (2014) 9−18.
[45] SIDEM, Selected References, 2014 [Online]. Available: http://www.sidem-desalination.com/en/main-references/case-studies/.
[46] SIDEM, Multiple Effect Distillation Using Mechanical Vapour Compression, 2014 [Online]. Available: http://www.sidem-desalination.com/en/Process/MED/MED-MVC/.
[47] AFFA, Economic and Technical Assessment of Desalination Technologies in Australia: With Particular Reference to National Action Plan Priority Regions, Department of Agriculture, Fisheries & Forestry − Australia, Turner, ACT, 2002.
[48] I. Moch, A 21st century study of global seawater reverse osmosis operating and capital costs, in: Proceeding of the IDA World Congress on Desalination and Water Reuse, in Manama, Bahrain, 2002.
[49] B. Rowlinson, D. Gunasekera, A. Troccoli, Potential role of renewable energy in water desalination in Australia, J. Renew. Sustain. Energy 4 (1) (2012) 013108.
[50] S. Miller, H. Shemer, R. Semiat, Energy and environmental issues in desalination, Desalination 366 (December 2015) 2−8.
[51] A. Ghobeity, A. Mitsos, Optimal design and operation of desalination systems : new challenges and recent advances, Curr. Opin. Chem. Eng. 6 (2014) 61−68.
[52] R. Semiat, Energy issues in desalination processes, Environ. Sci. Technol. 42 (22) (2008) 8193−8201.
[53] A.M. El-Nashar, Why use renewable energy for desalination, in: Desalination and Water Resources (DESWARE), Renewable Energy Systems and Desalination, vol. 1, EOLSS Publisher, 2010, pp. 202−215.
[54] B. Rahimi, J. May, A. Christ, K. Regenauer-Lieb, H.T. Chua, Thermo-economic analysis of two novel low grade sensible heat driven desalination processes, Desalination 365 (2015) 316−328.
[55] M. Elimelech, W.A. Phillip, The future of seawater desalination: energy, technology, and the environment, Science 333 (80) (2011) 712−718.
[56] S. Lattemann, T. Höpner, Environmental impact and impact assessment of seawater desalination, Desalination 220 (1−3) (March 2008) 1−15.
[57] A. Subramani, M. Badruzzaman, J. Oppenheimer, J.G. Jacangelo, Energy minimization strategies and renewable energy utilization for desalination: a review, Water Res. 45 (5) (February 2011) 1907−1920.
[58] C. Sommariva, H. Hogg, K. Callister, Environmental impact of seawater desalination: relations between improvement in efficiency and environmental impact, Desalination 167 (August 2004) 439−444.
[59] C. Fritzmann, J. Löwenberg, T. Wintgens, T. Melin, State-of-the-art of reverse osmosis desalination, Desalination 216 (1−3) (October 2007) 1−76.
[60] G.L. Meerganz von Medeazza, 'Direct' and socially-induced environmental impacts of desalination, Desalination 185 (1−3) (November 2005) 57−70.
[61] G. Raluy, L. Serra, J. Uche, Life cycle assessment of MSF, MED and RO desalination technologies, Energy 31 (13) (October 2006) 2361−2372.
[62] The World Bank, Renewable Energy Desalination an Emerging Solution to Close the Water Gap in the Middle East and North Africa, International Bank for Reconstruction and Development/The World Bank, Washington, D.C., 2012.
[63] T.-K. Liu, H.-Y. Sheu, C.-N. Tseng, Environmental impact assessment of seawater desalination plant under the framework of integrated coastal management, Desalination 326 (October 2013) 10−18.

CHAPTER 2

Low Grade Sensible Heat-Driven Distillation

2.1 Introduction to Low Grade Sensible Heat Sources

Typically, most of the thermal desalination units in seawater application or evaporation units in the mineral refining industry are powered by live steam. Live steam is one of the most valuable and expensive industrial energy resources and any process that can decrease the consumption of this precious resource, which in turn saves fuel consumption in the steam plant, will significantly reduce production cost and greenhouse gas emission and thus global warming.

In seawater applications, around 30–50% of the desalinated water cost is related to the energy cost of the desalination process [1], which is directly attributed to the live steam consumption in thermal desalination processes. This share underlines the importance of saving energy for reducing the water price and the desalination market in general. In industrial evaporation units, on the other hand, the price of freshwater production is not a direct issue. This is because the evaporation process subsumes under the main process and freshwater is not a direct outcome of the plant; so any expenses of the evaporation system will be reflected in the final product cost, which is different for different industries. For this purpose the solution is to monitor the amount of saved live steam from the optimized evaporation units, which can in turn be translated into tangible savings and economic benefits. With this purpose, the optimization of desalination and industrial distillation methods should be considered in the context of minimizing energy consumption [2] for which the importance of low grade sensible waste heat sources becomes evident.

Waste thermal energy has always been an important issue in the process industries. Management of waste heat resources is one of the important subjects in process plants. In general, industrial thermal energy can be divided into high grade and low grade types. High grade thermal energies are referred to as the heat that is viable to be recovered within the processes, whereas low grade heats are not and usually are rejected to the environment [3]. Temperature-wise, the threshold temperature range for low grade heat source is around 250°C [4]; therefore, any heat sources with temperatures lower than this can be classified as wastes. However, in this work, we restrict our consideration of those low grade heats that are between 60°C and 100°C. These types of low grade heat sources

Low Grade Heat Driven Multi-Effect Distillation and Desalination. http://dx.doi.org/10.1016/B978-0-12-805124-5.00002-4

are termed sensible heat sources if the waste heat medium is liquid, and is accompanied by a temperature drop during the heat transfer process. These kinds of heat sources, on account of their low temperatures, are suitable for conventional multi-effect distillation (MED) processes with a top brine temperature of around 70°C.

Low grade waste sensible heat sources are available in many industries, and for plants located in the coastal area, they can potentially be used for seawater desalination (e.g., [3,5–10]). In some other plants, these waste heat sources can be used for the evaporation processes of the plant for internal use, such as in alumina refinery plants (see Chapter 9) [11–13]. Management of these kinds of heat source has always been an important issue in the process industries. Desalination with low grade sensible heat sources is not limited to industrial waste heat streams. Low grade geothermal heat sources with a wellhead temperature lower than 100°C [14] can also be utilized for desalination purposes (e.g., [8,14–22]).

One of the main advantages of low grade sensible heat sources is related to carbon dioxide emission and global warming issues. If the required energy hails from a fossil fuel source then the freshwater production will contribute to carbon dioxide emission and consequently global warming. Low grade sensible heat sources such as waste heat from process plants and geothermal energy generate minimal greenhouse gasses.

Sensible waste heat sources and hot liquids generated from renewable energies are a promising sustainable energy source for desalination processes. Its true potential remains hitherto untapped. Referring to Fig. 2.1, for a reversible process, 1 kg of 75°C hot water

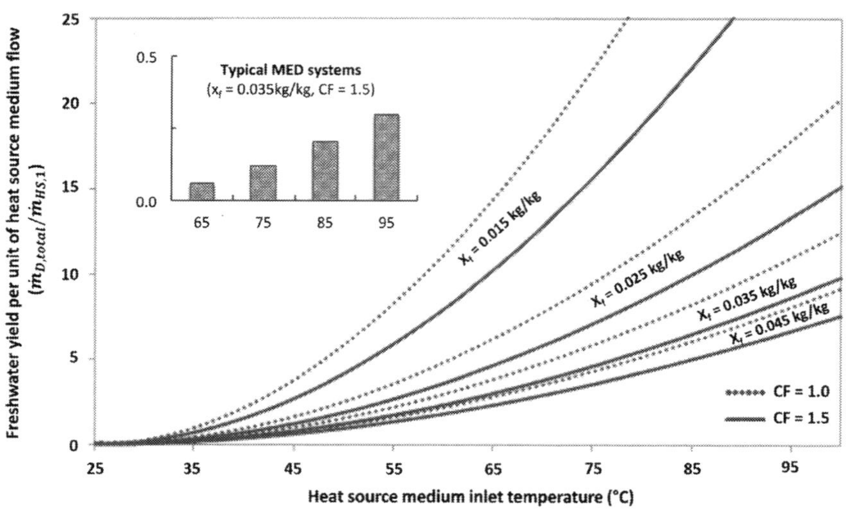

Figure 2.1

Thermodynamic limiting freshwater yield from a hot water stream at concentration factors (CF) of 1 and 1.5, respectively, with a cold sink temperature 25°C; Inset: typical yields of multi-effect distillation (MED) systems.

can generate 5.8 kg of pure water from seawater at a limiting zero recovery ratio. Conventional thermal distillation technologies like MED perform over an order of magnitude below this thermodynamic limit. One principal reason is that such thermal distillation technologies are almost invariably optimized for being steam driven. Sensible heat sources in contrast require specific approaches to unlock their potential, which are not addressed by these technologies and the common industrial benchmarks [9]. Consequently, the typical market approach of simply adapting a steam-driven design paradigm to sensible heat sources leads to thermodynamically inferior performance, as demonstrated in the inset of Fig. 2.1, which depicts the performance of market available MED systems intended for sensible heat sources. This figure considers both the thermodynamic least work of separation commonly used to benchmark reverse osmosis systems and the Carnot efficiency of converting the sensible heat source into work.

In this chapter, the MED as the conventional process (which is the most efficient process to couple with low grade sensible heat sources) and the preheated MED (P-MED) as an alternative solution for improving the conventional process, will be explained. Then, novel configurations—boosted MED (B-MED) [8,23], and flash-boosted MED (FB-MED) [2,24,25]—will be described as the superior technologies for low grade heat-driven desalination (evaporation) purposes as compared to the MED and P-MED processes.

2.2 Conventional Multi-Effect Distillation Process

It is known that thermal vapor compression multi-effect distillation (TVC-MED) has the highest performance among all steam-driven thermal desalination technologies [26–28]. It consumes less electrical energy and utilizes the thermal energy in an optimized way as compared to the other thermal desalination technologies. However, in the absence of medium-pressure steam and when the low grade sensible heat source (hot liquid medium) is the only available heat source, the conventional MED technology is a superior option.

MED is a proven desalination technology mimicking the natural water cycle. The separation is achieved by utilizing the differences in the volatilities of the fractions in a mixture. In 2013 MED accounted for 8% of the globally installed desalination capacity [29]. It is also applicable to a plethora of liquids, including industrial wastewater, spent liquor from the Bayer process (see Chapter 9), or otherwise contaminated waters. We shall concentrate herein on seawater desalination, but the general approach can be equally applied to any of the aforementioned applications.

In the temperature range of low grade sensible heat sources (<100°C), MED is ideal as its top brine temperature varies between 60°C and 75°C [2]. This process is more efficient

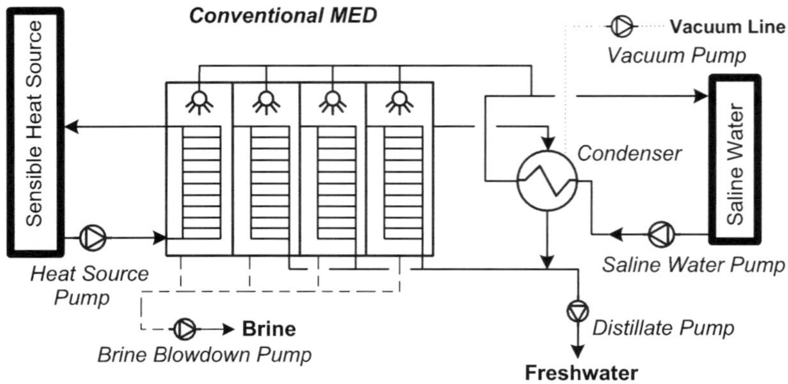

Figure 2.2
Schematic of a conventional parallel feed multi-effect distillation system coupled with a sensible heat source.

than other conventional thermal processes such as the multistage flash process. However, this process is still inefficient as the outgoing heat source temperature is still sufficiently high [9]. Fig. 2.2 shows a conventional MED system, coupled with a low grade sensible heat source. In the conventional MED process, the feed water is distributed onto the heat exchanger surfaces of the first effect, which is heated by the sensible heat source. The temperature drop occurs in the heat-source stream. The produced vapor then condenses in the next-effect heat exchanger and acts as a heat source for the feed water that is distributed in this effect; and the brine is purged. This process continues to the last effect. At the end, the corresponding produced vapor enters the condenser section and is condensed by the incoming saline feed water acting as a coolant.

2.3 Preheated Multi-Effect Distillation Process

In this configuration, the exiting heating medium is used to preheat the feed into the effects (Fig. 2.3). This offsets the sensible heating of the feed required to arrive at its boiling point. The major advantage of this system is its inherent simplicity, as only liquid–liquid heat exchangers are required. Also, some amount of energy recovery throughout the effects can be achieved for the heat added at those upstream effects. This design intrinsically constrains feed preheating up to the boiling point of the respective effect, as any evaporation in the liquid–liquid heat exchanger is to be avoided.

2.4 Boosted Multi-Effect Distillation Process

In cases of low grade sensible heat sources, it is far better to exploit the heat source as much as possible (insofar as operational and design limitations permit) to bring about maximal production rate [30]. The novel B-MED process, shown in Fig. 2.4, utilizes a

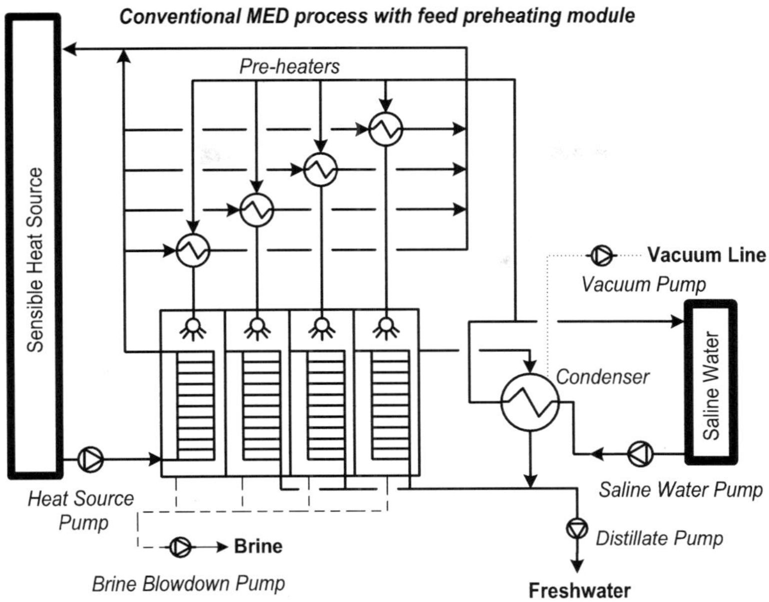

Figure 2.3
Schematic of a preheated multi-effect distillation system.

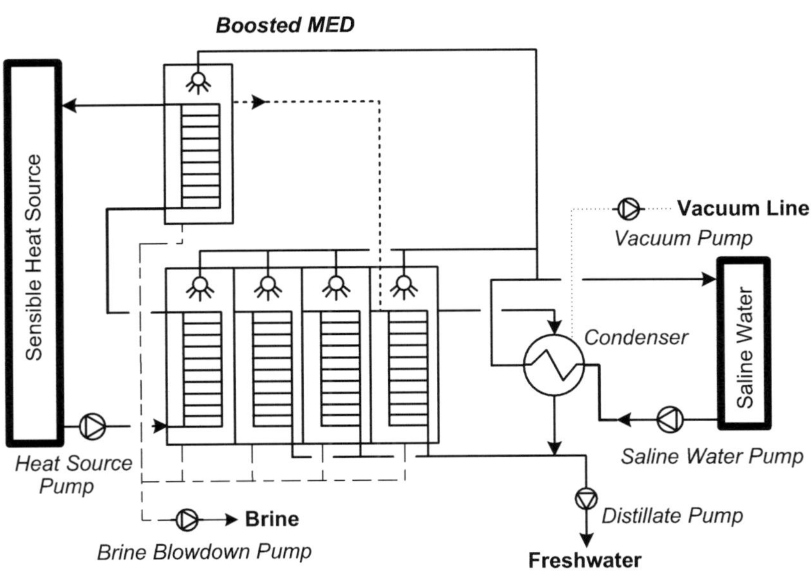

Figure 2.4
Schematic design of a boosted multi-effect distillation system.

booster unit, an evaporator unit that receives the heat source fluid that has already been used to provide the necessary thermal energy for the primary MED effects. The heat source fluid, which still contains a considerable amount of useful thermal energy, is used to evaporate more feed within the booster unit, which causes a higher temperature drop for the heat source medium as compared to the conventional MED process [8–10,23,31]. The vapor produced in the booster unit is then directed into an appropriate MED effect, supplementing the vapor and distillate production of the entire system. The B-MED process is particularly thermally efficient when utilizing lower temperature heat sources, achieving production rates greater than that of the optimized MED alternative utilizing the same heat source [2].

2.5 Flash-Boosted Multi-Effect Distillation Process

To further exploit waste sensible heat, an improved system, FB-MED, has been developed as shown in Fig. 2.5 [2,24,25]. The improvement is derived from the ability of the system to extract the maximal energy from the waste heat and transform it into heightened freshwater production in the primary MED plant. To this end, multiple flashing chambers are installed to efficiently transform the waste energy into valuable steam. Referring to Fig. 2.5, the outgoing heat source from the primary MED plant goes on to heat up the feed water via a liquid-to-liquid heat exchanger, which has been marginally preheated by the

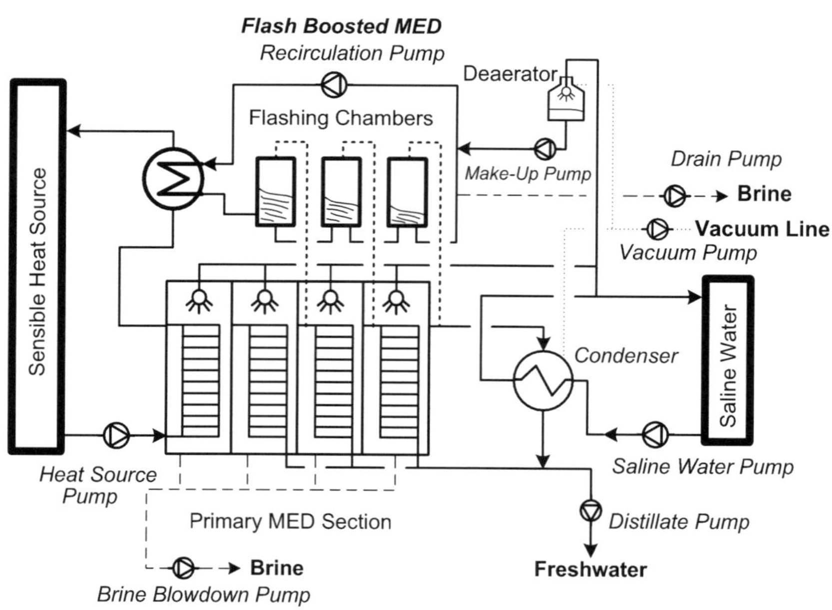

Figure 2.5
Schematic design of the flash-boosted multi-effect distillation (FB-MED) system, which depicts a 4-effect primary MED boosted by three flashing chambers.

outgoing brine from the last flashing chamber. The heightening of the feed water temperature via this approach is only limited by the practical temperature of approach of the liquid-to-liquid heat exchanger and top brine temperature and this scheme ensures that it comes sufficiently close to the outlet temperature of the heat source leaving the primary MED plant. The heated feed water then goes through a series of flashing chambers. Each flashing chamber produces supplementary vapor, which is directed to an appropriate primary MED effect according to the relevant pressure/temperature differences, thereby supplying more heat and increasing the amount of feed water that can be evaporated across the MED effect. The resulting effect of this vapor injection scheme is an increased distillate production across the series of primary MED effects as compared to all previous MED processes [2], which is made possible only by the addition of several pumps, flashing chambers, and a liquid–liquid heat exchanger [30].

FB-MED systems are not limited to the aforementioned configuration (Fig. 2.5). There are seven major configurations mentioned in the patent [24]. The other configurations are designed based on the combination of these seven major configurations. One of those is the flash-boosted thermal vapor compression multi-effect evaporation (FB-TVC-MEE) process, which is designed for mineral refinery plants (see Chapter 9).

References

[1] S. Miller, H. Shemer, R. Semiat, Energy and environmental issues in desalination, Desalination 366 (December 2015) 2–8.

[2] B. Rahimi, A. Christ, K. Regenauer-Lieb, H.T. Chua, A novel process for low grade heat driven desalination, Desalination 351 (October 2014) 202–212.

[3] Y. Ammar, S. Joyce, R. Norman, Y. Wang, A.P. Roskilly, Low grade thermal energy sources and uses from the process industry in the UK, Appl. Energy 89 (1) (January 2012) 3–20.

[4] Institution of Engineering and Technology, Profiting from Low-Grade Heat (The Watt Committee on Energy Report, No. 26), Institution of Engineering and Technology, London, 1994.

[5] Y. Ammar, H. Li, C. Walsh, P. Thornley, V. Sharifi, A.P. Roskilly, Desalination using low grade heat in the process industry: challenges and perspectives, Appl. Therm. Eng. 48 (December 2012) 446–457.

[6] H. Shih, Evaluating the technologies of thermal desalination using low-grade heat, Desalination 182 (1–3) (November 2005) 461–469.

[7] A. Ophir, F. Lokiec, Advanced MED process for most economical sea water desalination, Desalination 182 (1–3) (November 2005) 187–198.

[8] X. Wang, A. Christ, K. Regenauer-Lieb, K. Hooman, H.T. Chua, Low grade heat driven multi-effect distillation technology, Int. J. Heat Mass Transfer 54 (25–26) (December 2011) 5497–5503.

[9] A. Christ, K. Regenauer-Lieb, H.T. Chua, Thermodynamic optimisation of multi-effect-distillation driven by sensible heat sources, Desalination 336 (2014) 160–167.

[10] A. Christ, X. Wang, K. Regenauer-Lieb, H.T. Chua, Low grade waste heat driven desalination technology, Int. J. Simul. Multi. Design Optim. 5 (2013) 83–92.

[11] B. Rahimi, K. Regenauer-Lieb, H.T. Chua, E. Boom, S. Nicoli, S. Rosenberg, A novel low grade heat driven process to re-concentrate process liquor in alumina refineries, in: 10th International Alumina Quality Workshop (AQW) Conference, Perth, Australia, 19th–23rd April, 2015, pp. 327–336.

[12] B. Rahimi, K. Regenauer-Lieb, H.T. Chua, E. Boom, S. Nicoli, S. Rosenberg, A novel low grade heat driven process to re-concentrate process liquor in alumina refineries, Hydrometallurgy 2015 (2015) 327–336.

[13] B. Rahimi, K. Regenauer-Lieb, H.T. Chua, E. Boom, S. Nicoli, S. Rosenberg, A novel flash boosted evaporation process for alumina refineries, Appl. Therm. Eng. 94 (2016) 375−384.
[14] A.M. El-Nashar, Desalination with renewable energy—a review, in: Desalination and Water Resources (DESWARE), Renewable Energy Systems and Desalination, vol. 1, EOLSS Publisher, 2010, pp. 88−160.
[15] A. Husain, Renewable energy and desalination systems, in: Desalination and Water Resources (DESWARE), Renewable Energy Systems and Desalination, vol. 1, EOLSS Publisher, 2010, pp. 161−201.
[16] E. Mathioulakis, V. Belessiotis, E. Delyannis, Desalination by using alternative energy: review and state-of-the-art, Desalination 203 (1−3) (February 2007) 346−365.
[17] A.M. El-Nashar, Economics of small solar-assisted multiple-effect stack distillation plants, Desalination 130 (2000) 201−215.
[18] L. Garcia-Rodriguez, Seawater desalination driven by renewable energies : a review, Desalination 143 (2002) 103−113.
[19] K. Karytsas, V. Alexandrou, I. Boukis, The Kimolos geothermal desalination project, in: International Workshop on Possibilities of Geothermal Energy Development in the Aegean Islands Region, Milos Island, Greece, 2002, pp. 206−219.
[20] E. Barbier, Geothermal energy technology and current status : an overview, Renew. Sustain. Energy Rev. 6 (2002) 3−65.
[21] A.M. El-Nashar, Why use renewable energy for desalination, in: Desalination and Water Resources (DESWARE), Renewable Energy Systems and Desalination, vol. 1, EOLSS Publisher, 2010, pp. 202−215.
[22] V. Belessiotis, E. Delyannis, Renewable energy resources, in: Desalination and Water Resources (DESWARE), Renewable Energy Systems and Desalination, vol. 1, EOLSS Publisher, 2010, pp. 49−87.
[23] H.T. Chua, K. Regenauer-Lieb, X. Wang, A Desalination Plant, World Intellectual Property Organization, 2012. WO 2012/003525 A1.
[24] B. Rahimi, H.T. Chua, A. Christ, System and Method for Desalination, World Intellectual Property Organization, 2015. WO 2015/154142 A1.
[25] B. Rahimi, K. Regenauer-Lieb, H.T. Chua, A novel desalination design to better utilise low grade sensible waste heat resources, in: IDA World Congress 2015 on Desalination and Water Reuse, San Diego, US, 30-August to 4-September, 2015.
[26] H.T. El-Dessouky, H.M. Ettouney, Fundamentals of Salt Water Desalination, Elsevier Science B.V., 2002.
[27] M. Al-Shammiri, M. Safar, Multi-effect distillation plants: state of the art, Desalination 126 (1−3) (November 1999) 45−59.
[28] A.A. Al-Karaghouli, L.L. Kazmerski, Renewable energy opportunities in water desalination, in: Desalination, Trends and Technologies, InTech, 2011, pp. 149−184.
[29] T. Pankratz, IDA Desalination Yearbook 2012−2013, Media Analytics Ltd., Oxford, 2013.
[30] B. Rahimi, J. May, A. Christ, K. Regenauer-Lieb, H.T. Chua, Thermo-economic analysis of two novel low grade sensible heat driven desalination processes, Desalination 365 (2015) 316−328.
[31] A. Christ, K. Regenauer-Lieb, H.T. Chua, Development of an advanced low-grade heat driven multi effect distillation technology, in: IDA World Congress on Desalination and Water Reuse, 2013.

CHAPTER 3

Boosted Multi-Effect Distillation Pilot Plant

3.1 Introduction

As mentioned in Chapter 2, we introduced a novel boosted multi-effect distillation (B-MED) process to dovetail MED with low grade heat sources [1,2]. Compared with conventional MED, the boosted process features an increased capability to exploit low grade and waste heat sources.

3.2 Pilot Plant and Instrumentation

The pilot plant is realized via two serially connected single-effect rising film evaporation modules (Figs. 3.1 and 3.2). Alfa Laval plate heat exchangers are used for the evaporator and condenser sections, which consist of 16 0.4 mm M3-2/PO titanium plates for each section. The surface area per plate is 0.032 m^2 with a free channel width of 2.4 mm, resulting in a design pressure drop of 0.5 bar at the heating medium/cooling water side. Nitrile NBR gaskets are used for sealing. The feed water flow to the evaporator section is controlled by a spring loaded valve orifice, and maintained at an average feed flow of 0.145 kg/s for Module I and 0.132 kg/s for Module II.

A stainless steel mesh demister is placed between the evaporator and condenser sections to avoid the carryover of droplets. The evaporation units and the entire piping within the monitored area are well insulated with Aeroflex EPDM of 5–20 mm thickness depending on the local temperature gradient.

Two 4 kW Alfa Laval centrifugal pumps (CNL 40-40/200) are mounted on the rig for the cooling water supply. Due to the superior control characteristics however the pressure controlled Grundfos CRT16-4 (4 kW) VSP pumps installed in the laboratory have been used throughout the tests. For the freshwater extraction a 0.25 kW Desmi PWF-1525 centrifugal pump is employed per module. Brine extraction and removal of noncondensable gases are maintained by a combined brine/air ejector driven by the outgoing cooling water flow.

In preparation for a subsequent planned field test at an alumina refinery, the pilot plant is designed in accordance with the Australian mining regulations. To simplify transportation, the test rig can be split into two units, the size of two standard palettes, which can be forklifted.

Low Grade Heat Driven Multi-Effect Distillation and Desalination. http://dx.doi.org/10.1016/B978-0-12-805124-5.00003-6

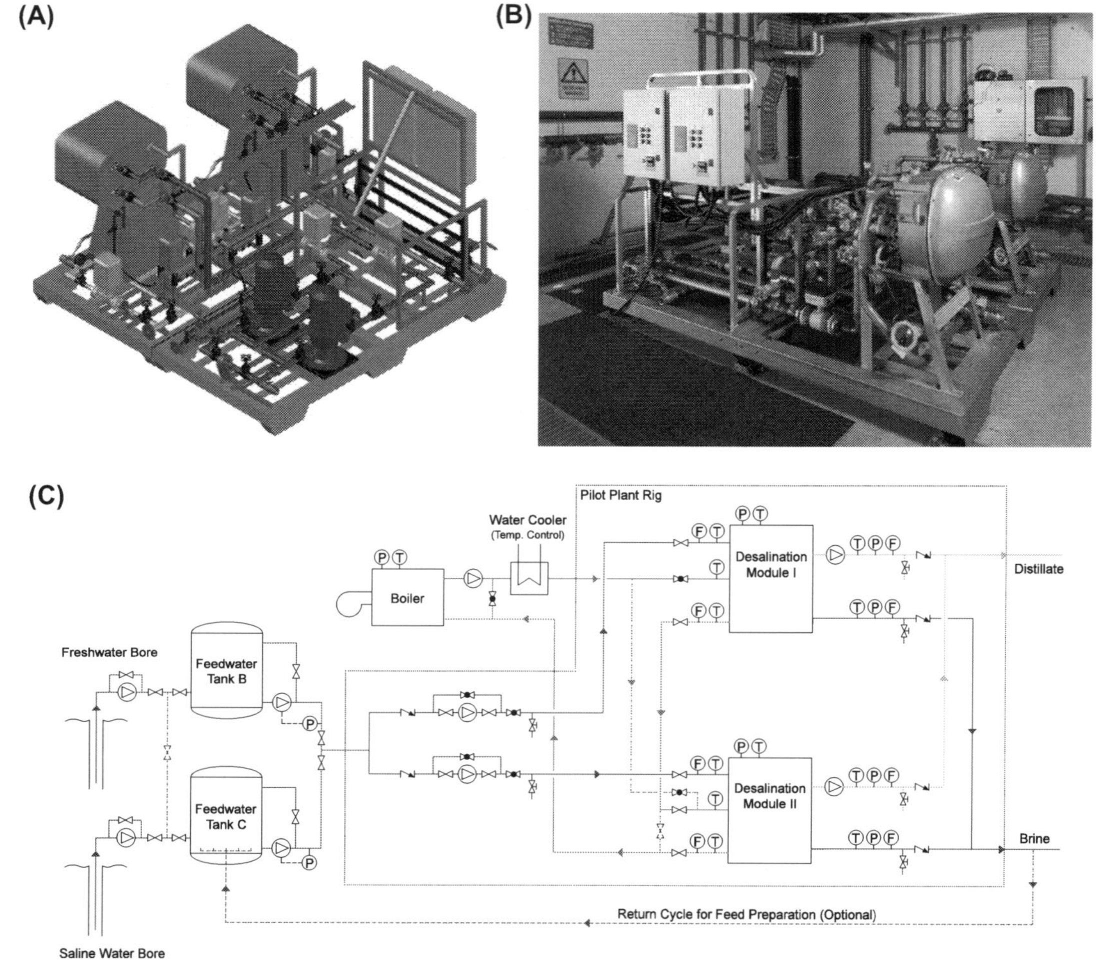

Figure 3.1
Technical drawing (A and C) and picture of the pilot plant (insulation removed) (B) [3].

The tests presented in this chapter had been undertaken at the National Center of Excellence in Desalination Australia (NCEDA) in Rockingham, Western Australia. The laboratory infrastructure includes two wells with a screened interval of 40–46 and 79–85 m below ground level, respectively, thereby providing feed water similar to seawater qualities (deep aquifer) and low-TDS (total dissolved solids) water (shallow aquifer). The detailed water analyses are shown in Section 3.4.2. The wellhead temperature varies seasonably between 18°C and 21°C. However to simulate common seawater temperatures 36 m^3 storage tanks were used for the preparation of cooling water batches of up to 35°C.

All tests had been carried out in a once-through arrangement in such a way that the produced freshwater and brine were rejected from the system. During the experiments, no antiscalant was added to the feed water and no separate feed pretreatment had been undertaken.

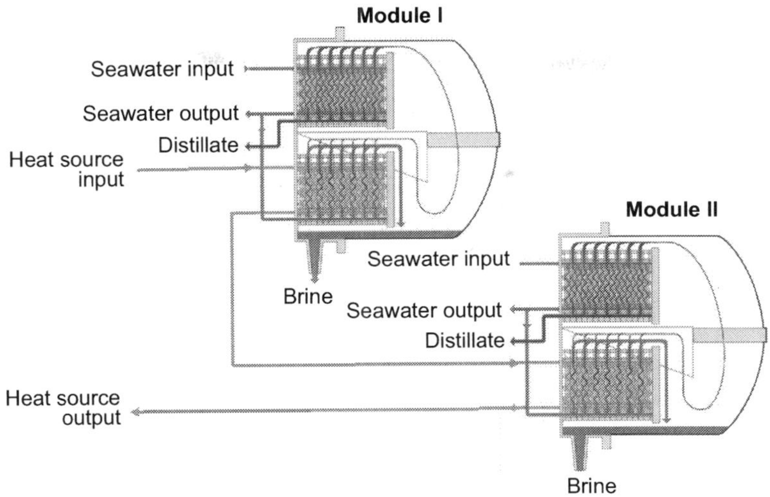

Figure 3.2
Schematic cross section of the evaporator units of the pilot plant [4]. *Adapted from ALFA LAVAL, Single Effect Freshwater Generator, Model JWP-16/26-C Series, in Alfa Laval Marine & Diesel Product Catalogue, Alfa Laval Corporate AB, 2003.*

The renewable energy or waste heat source is mimicked by using hot water provided by an ICI REX-9 90 kW boiler with a Riello 40-GS10 two-stage fan forced gas burner that provides heating medium input temperatures up to 93°C. Fluid circulation is provided by a Grundfos 32-80-180 centrifugal pump, with the heating medium flow rate being adjusted by a bypass connection upstream of the pilot plant.

To overcome limitations of the temperature control of the heating system, for heating medium temperatures $\leq$75°C, a water-cooled plate heat exchanger had been located upstream to the pilot plant to reject heat from the heating medium so as to maintain the desired heating medium temperature.

For the system analyses all incoming and outgoing process streams were monitored for flow rates and temperatures. Furthermore, temperature sensors were installed at the cooling water output and the vacuum space of each of the units. Also, pressure gauges were used to control the unit operation and determine the pressure drops at the heat exchangers. All instruments with signal output were logged in 30-s intervals. Table 3.1 summarizes the details of the instrumentation used.

3.3 Process Simulation and Validation

For the theoretical analysis of the system performance a steady state process simulation model has been developed [1,5,6]. The mathematical model will be explained in detail in Chapter 4.

Table 3.1: Instrumentation details [3].

Temperatures		
RTDs	Pt 100 1/10 DIN, ⌀ 3 mm	Calibrated (UKAS) at 20°C and 100°C ± 0.20°C (Assembly)
Flow Rates		
Cooling water and brine	Siemens SITRANS FM MAG 5100 W DN25 MAG 6000 Electromagnetic flowmeter	Calibrated (NIST) for water at 2 and 8 m^3/h ± 0.2% ± 1 mm/s
Distillate	Siemens SITRANS FC MASS 2100 DI6 Coriolis flowmeter	Calibrated (NIST) for water at 110, 260 and 490 kg/h: $\pm\sqrt{0.10^2 + \left(\frac{5}{\dot{m}}\right)^2}$ [%]
Pressures	Analog pressure gauges WIKA/Alfa Laval	[-]
Data Acquisition	Agilent 34972A LXI Data Acquisition Switch Unit Agilent 34901A Multiplexer Modules BenchLink DataLogger Pro Software	Calibrated (Agilent in-house) RTD[a]: ±0.06°C Flowmeter[a]: ±0.050% reading ±0.005% range

[a]Temperature coefficient negligible at lab conditions during tests.

For the validation of the simulation model against the actual results from the pilot plant a very good agreement has been achieved (Fig. 3.3). While at low productions the measurement accuracies of the input values of the simulation, because of the small temperature differentials, propagate to relatively large potential errors, all the simulated results are within a ±5% band of the actual experimental values. Considering the ideal nature of the simulation versus the real pilot plant, this deviation is well within the expected minor process losses.

3.4 Test Results

Test series had been undertaken with low and high saline waters (Table 3.2) over the temperature range common to sensible low grade heat sources, namely 68−93°C, and with cooling water temperatures ranging from 20°C to 35°C. Besides the temperature profiles of the two modules that are relevant to system design, the most crucial result is the extent of additional distillate produced by the downstream module in relation to the basic evaporation unit, as this will indicate the extent of additional pocket of vapor available for injection in the boosted MED configuration and therefore its ultimate viability. As an example, for the results as shown in Fig. 3.4, the second module produces 30 kg/s distillate, which is 50% of the production rate of the basic module (60 kg/s).

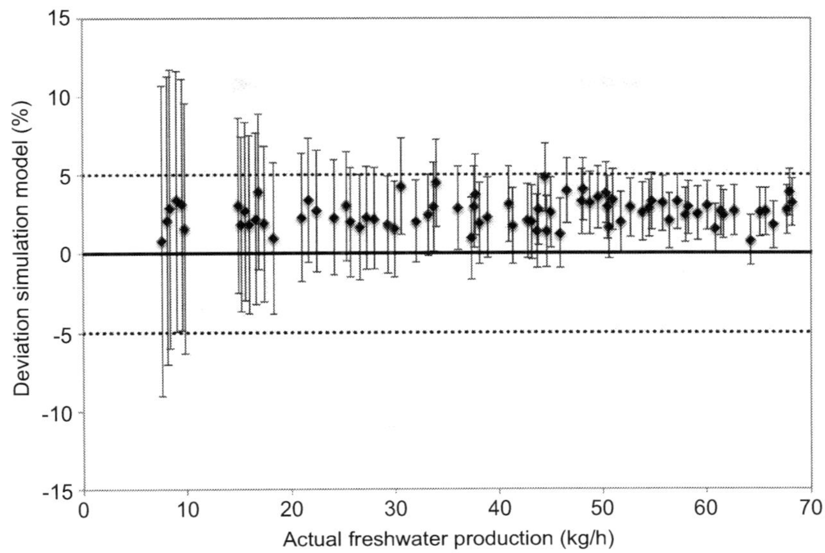

Figure 3.3
Deviation of the simulation results versus actual freshwater yield. Error bars indicate the propagated uncertainties of the input data to the simulation that stem from measurement inaccuracies [3].

3.4.1 Start-Up

Throughout the tests the pilot plant demonstrated a very stable operation (Fig. 3.4) as well as a fast response to altered heating medium input characteristics such as temperature. With the infrastructure capable of providing a stable heating medium input temperature to within $\pm 0.6°C$ and at constant flow rate, a very steady long-term operation can be achieved. This results in a good reproducibility of the data.

All the tests presented herein were undertaken over a time span of at least 60 min of stable operation; for visualization purposes, however, excerpts of 10 min intervals are used for the comparative figures. The start-up time of the pilot plant is dominated by the time span the boiler requires to reach the predetermined operational temperature, which is typically about 40 min (Fig. 3.4). The actual start-up time of the evaporator units is far shorter, with the necessary vacuum established within c.10 min.

Since no antiscalant had been used, after 120 h of operation surface scaling was visible on the feed side of the evaporator plates (Fig. 3.5). Also, solid scale manifestations of up to 30×90 mm occurred at a single area in the evaporator channels of both modules that typically receives lower feed flows and may encounter partial stagnation. However, no impact on the operational characteristic and the freshwater yield could be discerned from this degree of scaling.

Table 3.2: Water analysis [3].

Analyte	Unit	Limit of Testing	Sample Freshwater Bore	Sample Saline Water Bore	Sample Distillate Module I[a]	Sample Distillate Module II[a]
Sampled			**26/3/2013**	**26/3/2013**	**07/4/2014**	**07/4/2014**
Aluminum	mg/L	0.005	<0.005	<0.005	<0.005	0.005
Alkalinity, total	mg/L	1	216	146	<1	<1
Bicarbonate	mg/L	1	264	178	<1	<1
Boron	mg/L	0.02	0.06	3.2	<0.02	<0.02
Calcium	mg/L	0.1	48.6	468	<0.1	<0.1
Chloride	mg/L	1	99	15,300	1	<1
Carbonate	mg/L	1	<1	<1	<1	<1
Conductivity	uS/cm	2	—	—	<2	<2
Conductivity	mS/m	0.2	75.1	4360	<0.2	<0.2
Dissolved organic carbon	mg/L	1	<1.0	<1.0	<1.0	<1.0
Fluoride	mg/L	0.05	0.68	0.64	<0.05	<0.05
Hardness, total	mg/L	1	190	5600	<1	<1
Iron	mg/L	0.005	—	—	<0.005	<0.005
Iron(II)	mg/L	0.01	<0.01	<0.01	<0.01	<0.01
Iron(III)	mg/L	0.01	0.07	0.04	0.01	<0.01
Iron, total	mg/L	0.01	0.07	0.04	<0.01	<0.01
Magnesium	mg/L	0.1	22.7	1070	<0.1	<0.1
Manganese	mg/L	0.001	0.009	0.60	<0.001	<0.001
Nitrogen, ammonia	mg/L	0.01	0.08	1.1	<0.01	0.14
Nitrogen, nitrate	mg/L	0.01	<0.01	<0.01	<0.01	<0.01
pH		0.1	7.7	7.2	5.6	5.6
Phosphorus, sol. reactive	mg/L	0.01	0.28	0.07	<0.01	<0.01
Potassium	mg/L	0.1	2.1	258	<0.1	<0.1
Silica, molybdate reactive	mg/L	0.1	12	7.3	<0.1	<0.1
Silicon	mg/L	0.05	5.4	4.6	<0.05	<0.05
Sodium	mg/L	0.1	55.0	8550	<0.1	<0.1
Sulfur	mg/L	0.1	4.5	800	<0.1	<0.1
Sulfate	mg/L	1	15	2300	<1	<1
Sulfide	mg/L	0.01	0.01	<0.01	<0.01	<0.01
Total cations	mg/L	1	119	10,300	<1	<1
Total dissolved solids	mg/L	10	400	22,000	<10	<10
Total organic carbon	mg/L	1	<1.0	<1.0	<1.0	<1.0
Total suspended solids	mg/L	1	<1	18	<1	<1
Turbidity	NTU	0.5	<0.5	<0.5	<0.5	<0.5

[a]Samples taken at 78°C heating medium input temperature, 28°C cooling water temperature, and saline feed (saline water bore).

As a side note on the auxiliary power consumption of the pilot plant, the actual pressure drop encountered at the heating medium/cooling water side of the heat exchangers was experimentally confirmed as 0.5 bar at operational conditions, indicating that an efficient operation is viable. However, due to the deliberate selection of a cooling water-driven

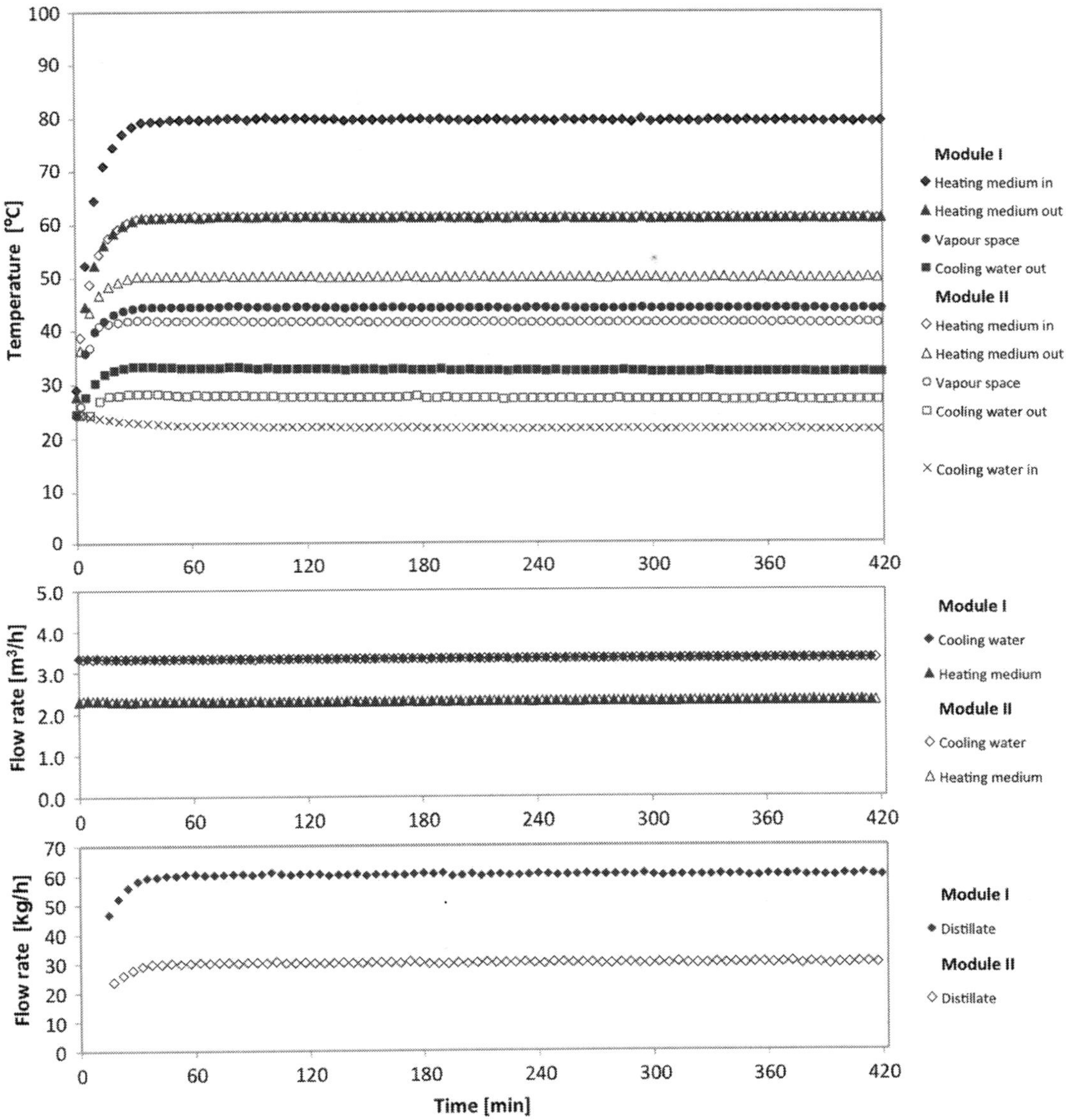

Figure 3.4
Start-up performance and operational stability of the pilot plant configuration [3].

ejector for brine and noncondensable gas removal to accommodate the anticipated harsh conditions of the subsequent field test at the alumina refinery plant, the cooling water cycle sustains a significant overall pressure drop of minimum 3.5 bar. Coupled with the lack of additional effects in the pilot plant configuration, this results in an unrepresentative high auxiliary power consumptions in excess of 30 kWh/m^3 for the current test setup.

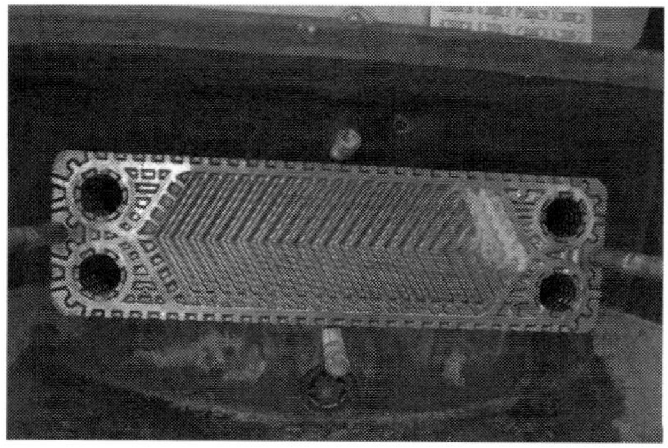

Figure 3.5
Evaporator plate surface condition after 120 h of operation without using any antiscalant.

3.4.2 Product Quality

The produced distillate has been analyzed by an independent analytical laboratory (ChemCentre, WA), which demonstrates exemplary distillate quality for both modules with all relevant parameters being at or below the limit of testing (Table 3.2). Frequent conductivity measurements corroborated this distillate quality over the entire test period. Since no significant difference in the water quality from each of the two units is recognizable, no negative influence of the booster module on the water quality of the basic MED is to be expected.

3.4.3 Effect of Heat Source and Cooling Water Temperatures

The heating medium input temperature is a major factor for the production rate of the distillation system since it determines the available energy input into the system. Fig. 3.6 depicts the results of test series at 93°C, 85°C, 75°C, and 68°C with all other operating parameters being kept constant. The only exception being the 93°C test series, since in order to arrive at this elevated temperature the heating medium flow rate had to be reduced as highlighted in the chart. Generally owing to the larger available temperature gradient, a higher heating medium temperature engenders an increased production rate. However when the energy intake into the system is kept constant by a corresponding reduction in the heating medium flow rate despite an increase in the temperature, that being the case at 93°C, which is derived from the 85°C test, the production rate is maintained at a level equivalent to the 85°C test.

Referring to the temperature profiles in Fig. 3.6, however, it can be seen that the utilized temperature gradient in both modules gradually decrease with the reduced heating medium temperature, with the downstream module efficiently utilizing the output stream of the first module. The vapor space temperature of the second module consistently levels at a

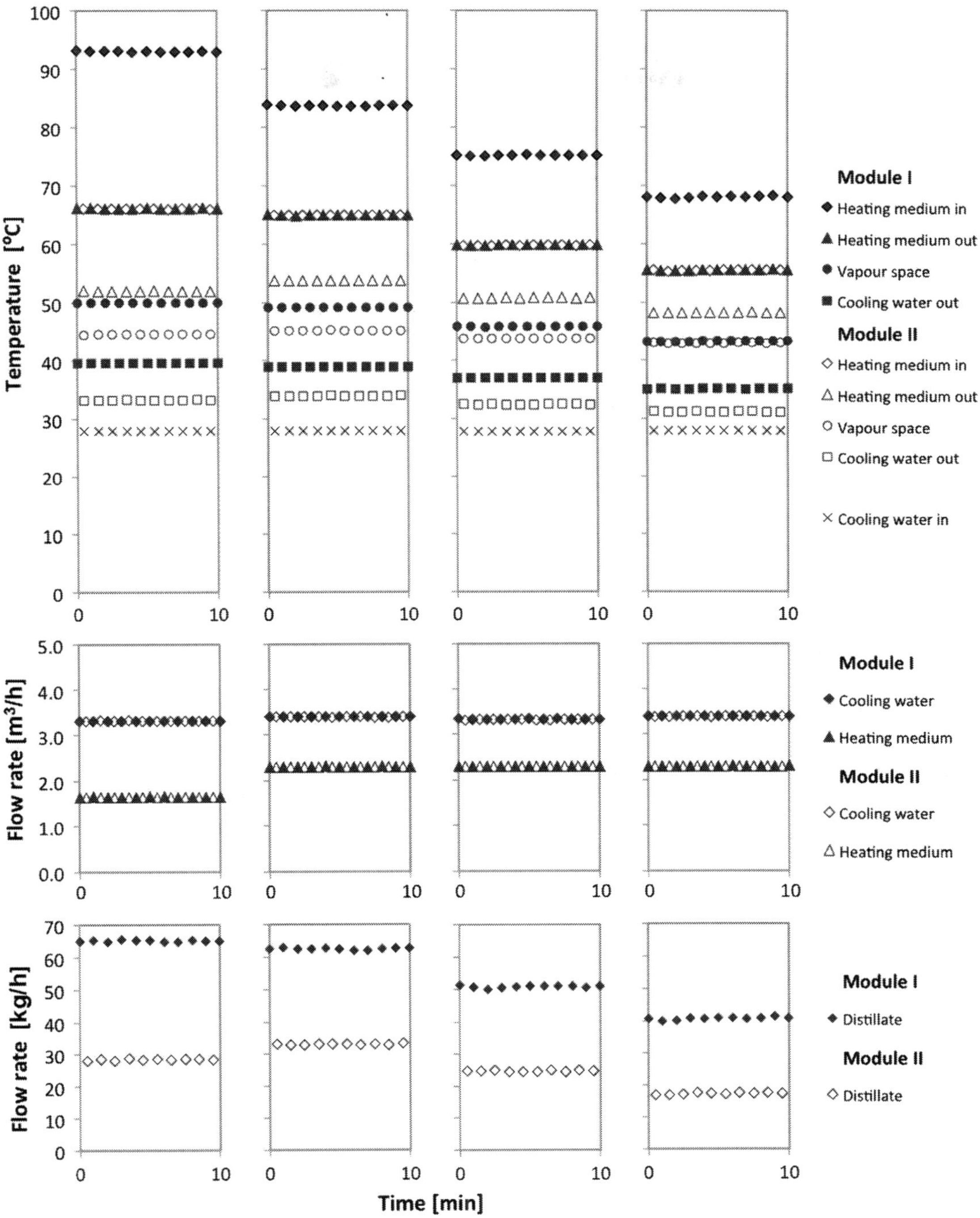

Figure 3.6
Effect of the heat source input temperature on the temperature distribution and the pilot plant performance [3].

temperature of 43–45°C throughout the tests, thereby demonstrating sufficient vapor pressure for a contemplated injection in a typical downstream effect of an MED plant [7] as in the case of the boosted MED process.

The hotter first module maintains vapor space temperatures of 50°C at 93°C heating medium temperature, but approaches the temperature level of the second unit where a reduced input temperature enjoins a lower boiling point. The cooling water output temperatures are a function of the energy intake into the system, with reduced heating medium temperatures giving rise to lesser cooling requirements. In the context of the additional vapor production for the boosted MED, this implies that the produced vapor flow from the booster module increases with the heating medium temperature. While this results in overall higher production rates, the relative effect of the booster typically declines with increasing heat source temperature.

The unit was tested at cooling water temperatures of 20°C, 28°C, and 35°C. Fig. 3.7 displays these results at an exemplary heating medium input temperature of 75°C. While a higher cooling water temperature necessarily lowers the sensible heating requirement of the feed to reach the boiling point, this benefit is negated by the impact on the available temperature differentials for the modules. As a matter of course, higher cooling water input temperatures lead to increased cooling water output temperatures, which limits the minimal boiling point temperature in the modules. Consequently, the potential temperature differential accessible by the plant is reduced, thereby resulting in dropping production rate. For multi-effect systems, this is exacerbated by a reduction in the number of effects, thereby directly militating against the maximal freshwater yield of the system.

The vapor saturation temperature in the downstream module, however, rises with increasing cooling water temperature, thereby maintaining sufficient pressure for the contemplated injection in the MED process, as in the case of the boosted MED scheme. The actual production rate of this module over the surveyed range is resilient against changes in the temperature of the cooling water temperature and remains substantial for a contemplated injection.

3.4.4 Effect of Feed Water Salinity

Experiments were also conducted at two disparate salinity levels over the entire temperature range. Fig. 3.8 illustrates the performance at a TDS concentration of 400 and 22,000 mg/L, respectively, demonstrating that the only actual recognizable deviation is found in the vapor space temperature of the modules as resulting from the increasing boiling point elevation of the feed at higher salinity levels. Consequently, besides a potentially higher propensity to scaling at increased feed water salinities, no significant negative impact on the performance of the boosting is expected within the common salinity range.

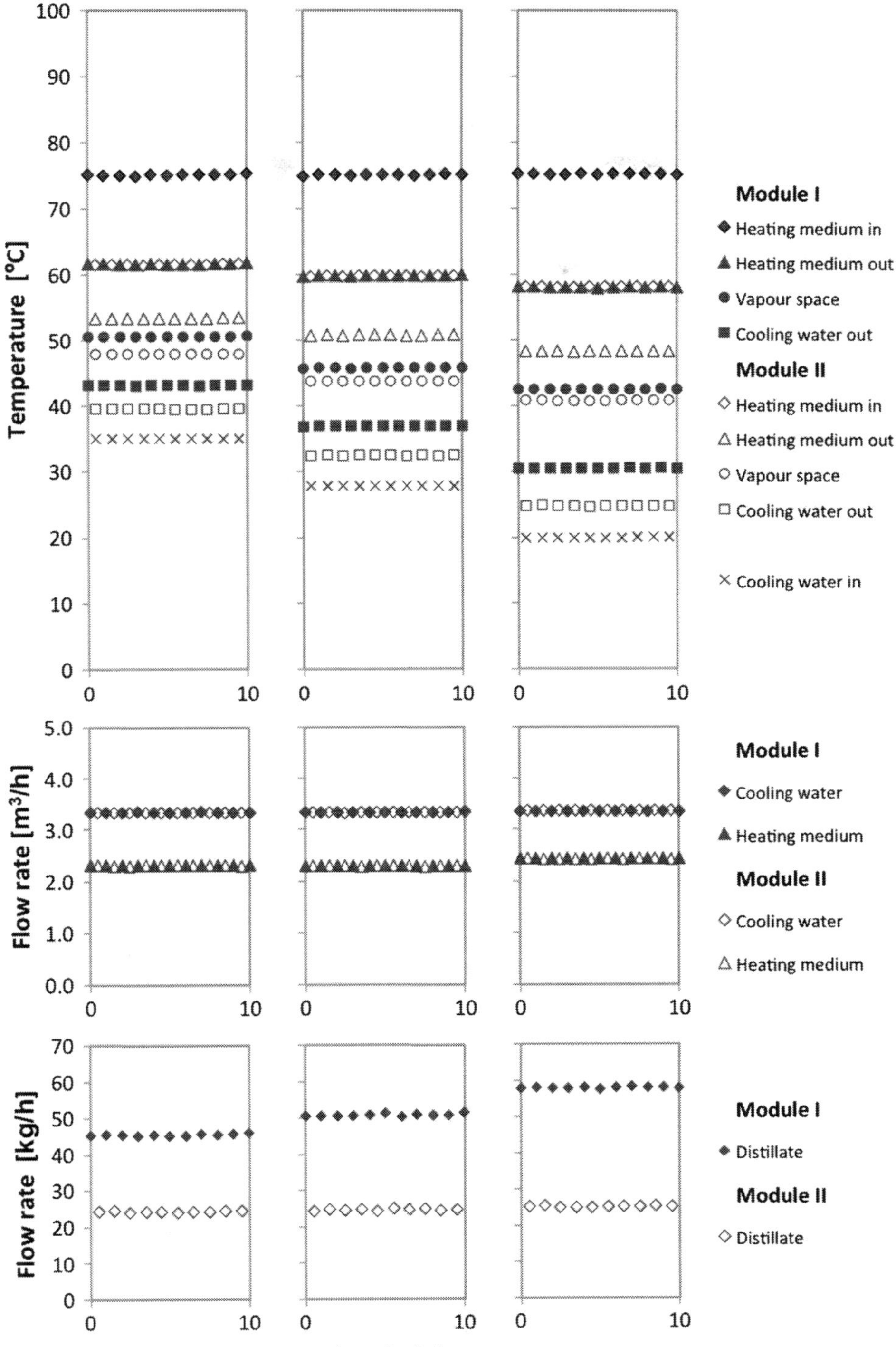

Figure 3.7
Effect of the cooling water temperature on the temperature distribution and the pilot plant performance [3].

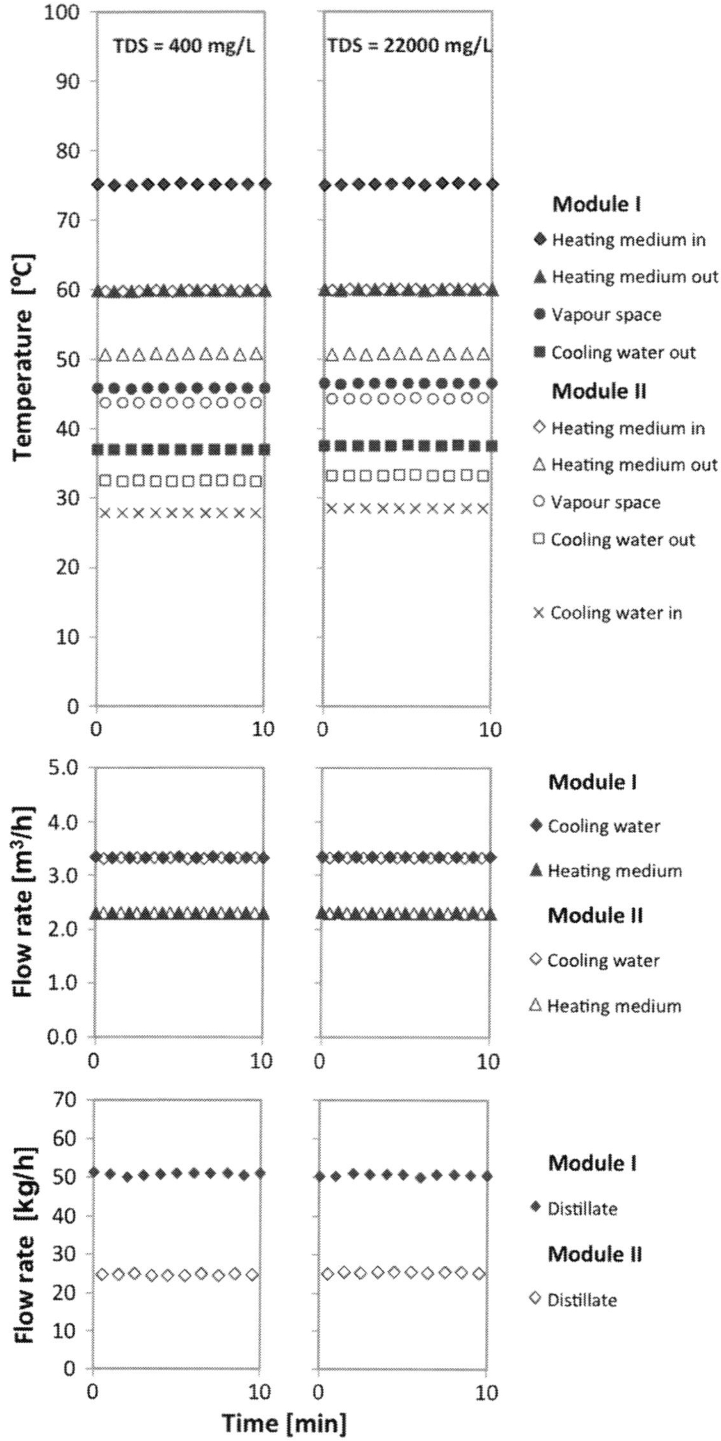

Figure 3.8

Effect of the feed water salinity on the temperature distribution and the pilot plant performance [3].

3.4.5 Potential of Boosting

In terms of the freshwater production results over the entire range of application, the downstream module is capable of producing a substantial amount of vapor at all tested conditions. Referring to Fig. 3.9, the additional vapor generated by the second module and therefore available for the contemplated boost to a basic MED process ranges from 33% to 57%. The lowest heating medium temperature tested with the unit was 57°C at a cooling medium temperature of 25°C, whereby an additional 27% freshwater production was achieved by Module II.

This verifies the fundamental basis of the boosted MED layout, which relies on the substantial amount of steam generated by a downstream evaporator. We qualify that in an actual boosted MED scheme, in which both the primary MED plant and steam injection from the booster are optimized, the final gain in freshwater production resulting from the booster module will be somewhat lower than 33–57% as reported herein, since in the final boosted configuration both process streams are reused in a different number of MED effects. Depending on the actual plant layout and operational conditions, the substantial additional amount of steam from the booster as reported herein translates to about 15–40% higher freshwater productions of an entire optimized MED plant [5].

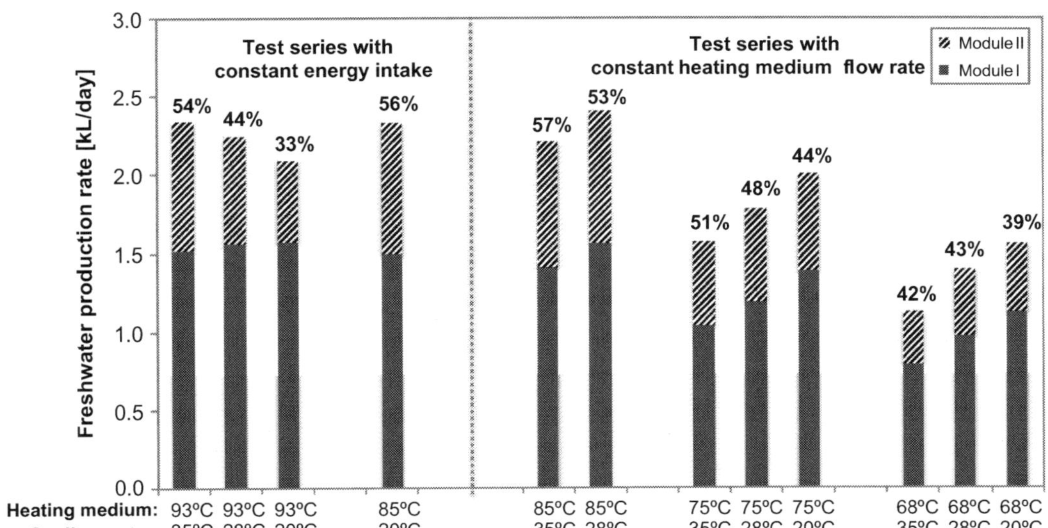

Figure 3.9

Freshwater production rate of both modules at various temperatures. The percentage above the columns indicates the additional production of Module II, which is potentially available for a vapor boost of the basic multi-effect distillation scheme [3].

3.5 Conclusion

The novel boosted MED design overcomes the thermodynamic limitations of conventional MED designs as applied to sensible heat sources. By accessing a larger heating medium temperature interval, significantly higher freshwater production rates can be achieved as compared to common systems. This augurs well for a significant potential for waste heat and renewable energy-driven desalination. The 2.5 m^3/day pilot plant as tested at the NCEDA has successfully demonstrated the key features of this design concept. By employing an evaporator module, which mimics the booster in a boosted MED scheme, and powered by the heating medium output stream of an identical upstream module, which represents the first effect of a primary MED plant, we have successfully demonstrated the following:

- A substantial amount of additional vapor, ranging from 33% to 57% of the production rate of the basic module, can be produced by the downstream evaporator.
- The boiling point of the additional vapor is sufficiently high to match downstream effects of common MED systems, thereby confirming that the contemplated steam injection is possible.
- The distillate produced from both modules is of very high quality. All relevant parameters are on or below the limit of testing.
- While the heating medium input temperature and the cooling water temperature exert significant influence on the overall performance of the system, no significant effect can be discerned from the feed water TDS concentration over the tested application range.
- The process simulation model used for system design and optimization has been experimentally proven to be congruent with the pilot plant performance.

This validates the principal design hypothesis of the boosted MED scheme and proves the general viability of the concept. The additional vapor generated in substantial quantity underpins the potential of the system. A more comprehensive analysis taking into account all economic parameters for a contemplated field application will be undertaken in Chapter 8.

References

[1] X. Wang, A. Christ, K. Regenauer-Lieb, K. Hooman, H.T. Chua, Low grade heat driven multi-effect distillation technology, Int. J. Heat Mass Transfer 54 (25–26) (December 2011) 5497–5503.

[2] H.T. Chua, K. Regenauer-Lieb, X. Wang, A Desalination Plant, World Intellectual Property Organization, 2012. WO 2012/003525 A1.

[3] A. Christ, K. Regenauer-Lieb, H.T. Chua, Application of the boosted MED process for low-grade heat sources—a pilot plant, Desalination 366 (2015) 47–58.

[4] ALFA LAVAL, Single Effect Freshwater Generator, Model JWP-16/26-C Series, in Alfa Laval Marine & Diesel Product Catalogue, Alfa Laval Corporate AB, 2003.

[5] A. Christ, K. Regenauer-Lieb, H.T. Chua, Thermodynamic optimisation of multi-effect-distillation driven by sensible heat sources, Desalination 336 (2014) 160–167.
[6] A. Christ, X. Wang, K. Regenauer-Lieb, H.T. Chua, Low-grade waste heat driven desalination technology, Int. J. Simul. Multi. Design Optim. 5 (February 2014) A02.
[7] L. Awerbuch, in: Understanding of Thermal Distillation Desalination Processes, IDA Academy, Singapore, 2012.

CHAPTER 4

Mathematical Simulation

4.1 Introduction

Our mathematical model is developed based on the steady state mass and energy balances coupled with the heat transfer equations for each individual element—the effects, condenser, heat exchangers, flashing chambers—and conjoining them with the ratio between the mass of feed water to that of produced freshwater. In the following sections we will provide a detailed explanation for each section of the various configurations (conventional multi-effect distillation (MED), preheated (P)-MED, boosted (B)-MED, and flash-boosted (FB)-MED). For a better understanding of what happens in each heat transfer unit of the proposed configuration, the relevant temperature–energy profiles will be plotted. *UA* values, being the essential design parameter for heat transfer units such as evaporators, condensers, and heat exchangers, will be explained as well.

Our mathematical model has been calibrated with the single-effect distillation (SED) plant performance data available from the Alfa Laval Marine & Diesel product catalog [1]. These data are tabulated in Table 4.1. The calibrated model is then used to demonstrate the amelioration achieved by our proposed designs such as P-MED, B-MED, and FB-MED processes. This enables us to validate the fundamental basis of our proposed design using the Alfa Laval SED plant data [1]. Our proposed designs are then applied to MED processes instead of SED processes under different operation conditions commonly encountered in field operations.

Fig. 4.1 depicts a comparison between the predicted freshwater yield from the simulation and the actual freshwater yield from SEDs reported in the Alfa Laval single effect freshwater generators catalog [1]. In total, 48 cases in the catalog stemming from different heat source temperatures, heat source flow rates, and freshwater generator capacities are studied. It is evident that the predicted data match the actual data very well for all the 48 cases, with the prediction error being within $\pm 5\%$ for most cases.

4.2 Mathematical Simulation Method

As with other thermal processes, mathematical simulation of these processes includes mass, salinity, and energy balance equations. Relevant equations in an equation system are solved to calculate important parameters such as production rate, power consumption, waste heat performance ratio, *UA* values, temperature gradients across each effect, and other

Low Grade Heat Driven Multi-Effect Distillation and Desalination. http://dx.doi.org/10.1016/B978-0-12-805124-5.00004-8

Table 4.1: Single-effect distillation plant performance data available from the Alfa Laval Marine & Diesel product catalog ($T_{C,in}$ = 32°C, X_C = 35,000 ppm) [1].

$\dot{m}_{D,total}$ m³/d	$T_{HS,1,in}$ °C	$T_{HS,1,out}$ °C	$\dot{m}_{HS,1}$ kg/s	$\dot{Q}_{HS,1}$ kW	$\dot{m}_C$ kg/s
1	65	55.2	0.83	34	1.16
	68	57.3	0.69	31	1.16
	70	58.1	0.72	36	1.16
	75	59.9	0.55	35	1.16
	78	61.3	0.50	35	1.16
	80	63.8	0.47	32	1.16
	85	68.8	0.47	32	1.16
	90	70.5	0.42	34	1.16
2	65	54.3	1.38	62	2.18
	68	56.1	1.24	62	2.18
	70	57.0	1.14	62	2.18
	75	59.7	0.97	62	2.18
	78	61.6	0.92	63	2.18
	80	63.0	0.88	63	2.18
	85	66.9	0.83	63	2.18
	90	68.8	0.72	64	1.16
3	65	53.5	2.22	107	3.32
	68	56.6	1.94	93	3.32
	70	53.4	1.58	110	3.32
	75	58.3	1.50	105	3.32
	78	59.9	1.38	105	3.32
	80	63.3	1.33	93	2.18
	85	66.7	1.33	102	2.18
	90	65.9	0.97	98	2.18
4	65	54.4	2.76	123	4.34
	68	56.1	2.48	124	4.34
	70	57.0	2.66	145	4.34
	75	61.3	2.21	127	3.32
	78	61.9	2.04	138	3.32
	80	62.5	1.71	126	3.32
	85	66.4	1.60	125	3.32
	90	66.8	1.36	132	3.32
5	65	57.7	5.00	153	6.24
	68	57.9	3.61	153	7.09
	70	57.5	2.91	153	5.67
	75	60.6	2.76	167	4.34
	78	61.2	2.27	160	4.34
	80	62.9	2.21	159	4.34
	85	66.0	2.04	163	3.32
	90	68.2	1.77	162	3.32
7	65	57.3	6.65	215	8.51
	68	59.6	6.07	214	6.24
	70	60.8	5.54	214	6.24
	75	60.9	3.60	213	7.09

Table 4.1: Single-effect distillation plant performance data available from the Alfa Laval Marine & Diesel product catalog ($T_{C,in}$ = 32°C, X_C = 35,000 ppm) [1].—cont'd

$\dot{m}_{D,total}$ m³/d	$T_{HS,1,in}$ °C	$T_{HS,1,out}$ °C	$\dot{m}_{HS,1}$ kg/s	$\dot{Q}_{HS,1}$ kW	$\dot{m}_C$ kg/s
	78	60.7	2.90	211	5.67
	80	62.8	2.91	210	5.67
	85	66.7	2.77	213	4.34
	90	68.4	2.49	226	4.34
10	65	57.2	9.37	307	10.21
	68	59.5	8.57	306	10.21
	70	60.9	7.98	305	8.51
	75	64.5	6.89	304	8.51
	78	66.7	6.36	302	6.24
	80	68.3	6.11	300	6.24
	85	65.3	3.59	297	7.09
	90	70.4	3.60	296	7.09
15	65	57.1	13.86	460	15.88
	68	60.6	14.64	455	13.89
	70	60.8	11.85	458	13.89
	75	64.4	10.22	455	10.21
	78	66.6	9.44	452	10.21
	80	67.9	8.86	450	10.21
	85	71.8	8.05	446	8.51
	90	71.7	5.79	445	8.51

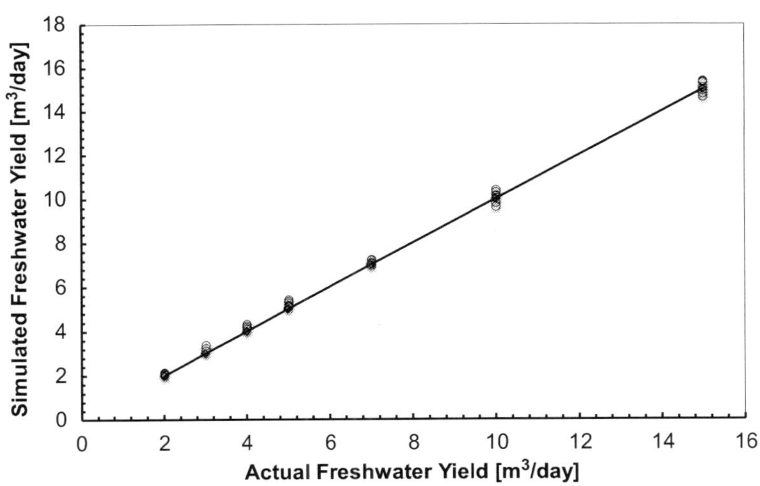

Figure 4.1
A comparison between our predicted freshwater yield and actual freshwater yield derived from the Alfa Laval single effect freshwater generators [1].

parameters, all of which are useful for a thermoeconomic evaluation. In Section 4.2.6, the relevant solving procedure flowchart of the proposed processes for seawater application is shown. A generalized reduced gradient (GRG) method [2] is used for solving the equation system, by imposing the boundary conditions from the laws of thermodynamics and appropriate operational and technical constraints. The outcome of this mathematical model is used in the economic analysis, which is explained in detail in Chapter 7.

The assumptions that are considered for the mathematical simulation are:

1. Steady state process.
2. Negligible external heat loss across each effect and flashing chamber.
3. Constant inlet feed water temperature and salinity.
4. Negligible internal pressure/temperature losses of the vapor flow in the demister, transmission lines, and condensation inside the tubes.
5. All condensation processes happen under constant pressure.
6. The produced water is pure water.
7. All primary MED effects have an equal and constant recovery ratio.
8. A 500 Pa pressure difference is considered for injecting vapor from the flashing chambers to the relevant MED effect [3].
9. A constant 2.5°C temperature difference is considered inside each primary MED effect, between the condensed vapor temperature (heat source) and the outlet brine or concentrated feed temperature [4,5].

For the mathematical simulation, the mass, salinity, and energy balance equations are written for each part of all the proposed configurations, which will be explained in the following sections. These parts include the first effect of the primary MED, the second to the last effect of the primary MED (these include the injected effects for the B-MED and FB-MED processes), the condenser, the preheaters (for P-MED process), the booster of the B-MED process, the brine heater, the deaerator and the battery of flashing chambers of the FB-MED process.

For the properties of water, steam, and seawater, the REFPROP package (REFerence fluid PROPerties) developed by the National Institute of Standards and Technology (NIST) [6], and reference [7] have been used, respectively. The properties of NaOH (see Chapter 9) are consulted from [8–13].

4.2.1 Conventional Multi-Effect Distillation Process

As shown in Fig. 4.2, the sensible heat-driven conventional MED includes three main sections:

1. The first effect, which is driven with hot liquid as its heat source.
2. The second to the last effect, which are steam-driven evaporators.
3. The last condenser, which is cooled by saline water (for seawater application).

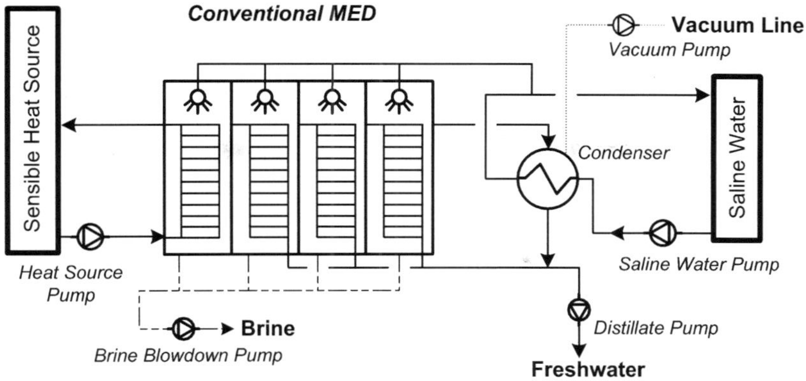

Figure 4.2
Schematic design of a four-effect conventional multi-effect distillation process coupled with a sensible heat source.

The mass and salinity balance for the evaporation side of all n-MED effects (the first to the last) are written as ($k \in \{1,\ldots,n\}$):

$$\dot{m}_{F,k} = \dot{m}_{B,k} + \dot{m}_{V,k} \tag{4.1}$$

$$\dot{m}_{F,k} \cdot X_{F,k} = \dot{m}_{B,k} \cdot X_{B,k} \tag{4.2}$$

With reference to the definition of feed-to-vapor ratio (R) and recovery factor (RF) which is defined as

$$R = \frac{1}{\mathrm{RF}} = \frac{\dot{m}_F}{\dot{m}_V} \tag{4.3}$$

the mass flow rate and salinity of the outlet concentrated feed (which is also known as brine in seawater application) can be written as

$$\dot{m}_{B,k} = (R-1) \cdot \dot{m}_{V,k} = \left(\frac{R-1}{R}\right) \cdot \dot{m}_{F,k} \tag{4.4}$$

and

$$X_{B,k} = \left(\frac{R}{R-1}\right) \cdot X_{F,k} \tag{4.5}$$

The energy balance for the MED sections should be written separately for the first effect, from the second effect to the last effect, and from the condenser section, all of which are explained next.

4.2.1.1 First Effect

The inlets and outlets of the first effect are shown in Fig. 4.3. As explained in Chapter 2, the heat source medium enters this effect and after releasing its sensible thermal energy,

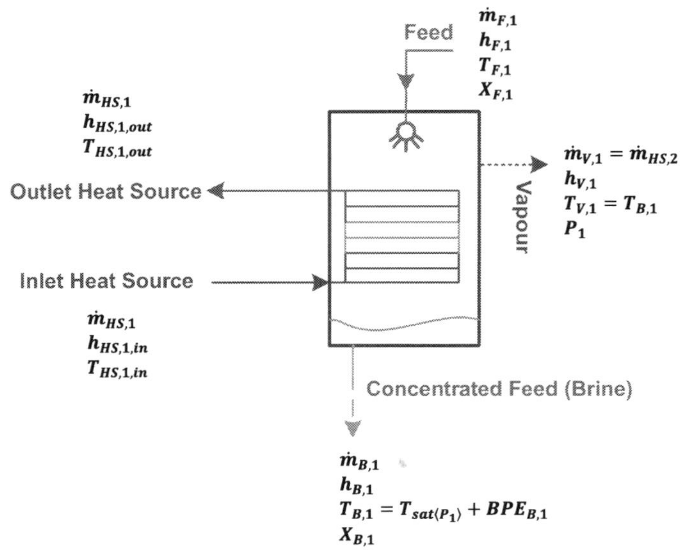

Figure 4.3
Schematic design of the first effect.

its temperature drops and it then leaves the effect. The released energy is transferred to the inlet feed stream, which is distributed from the top onto the hot pipes or plates (depending on the type of evaporator). The outlet of the feed stream side is a concentrated feed stream (called brine in seawater application), which is purged from the bottom. The vapor, which is directed to the second effect, is then used as the heat source of that effect.

4.2.1.1.1 Energy Balance

The energy balance for the first effect according to Fig. 4.3 is written as

$$\dot{m}_{F,1} \cdot h_{F,1} + \dot{m}_{\text{HS},1} \cdot h_{\text{HS},1,\text{in}} = \dot{m}_{\text{HS},1} \cdot h_{\text{HS},1,\text{out}} + \dot{m}_{V,1} \cdot h_{V,1} + \dot{m}_{B,1} \cdot h_{B,1} \tag{4.6}$$

$h_{F,1}$ is the enthalpy of the feed water, which is a function of its temperature and salinity (see Appendix A for seawater application), namely

$$h_{F,1} = h_{f_F \langle T_{F,1}, X_{F,1} \rangle} \tag{4.6.1}$$

$h_{\text{HS},1,\text{in}}$ and $h_{\text{HS},1,\text{out}}$ are the heat source medium inlet and outlet enthalpies of the first effect, respectively. In the case of using condensed steam (pure water) as the heat source, these enthalpies are only a function of temperature (if they are saturated) and can be found from the steam tables or relevant software such as REFPROP [6]; otherwise they are a function of both temperature and salinity of their medium and should be calculated based on their relevant equations.

$$h_{\text{HS},1,\text{in}} = h_{f_{\text{HS}} \langle T_{\text{HS},1,\text{in}}, X_{\text{HS},1} \rangle} \tag{4.6.2}$$

$$h_{\text{HS},1,\text{out}} = h_{f_{\text{HS}} \langle T_{\text{HS},1,\text{out}}, X_{\text{HS},1} \rangle} \tag{4.6.3}$$

$h_{V,1}$ is the enthalpy of the produced vapor, Eq. (4.6.4), which is slightly superheated in seawater applications and completely superheated in process liquor applications such as in alumina refinery plants (see Chapter 9). In the latter situation, the produced vapor pressure is equal to the effect's pressure (P_1), but its temperature is significantly more than the saturated temperature at P_1, by the boiling point elevation (BPE) $\left(T_{V,1} = T_{B,1} = T_{\text{sat}\langle P_1 \rangle} + BPE_{B,1} > T_{\text{sat}\langle P_1 \rangle}\right)$. In seawater applications the BPE is not more than 1°C (see Appendix B), therefore the degree of superheat is small and negligible, but in contrast in process liquor applications such as in alumina or other mineral refinery processes, the BPE (which is a function of temperature and concentration; see Appendix B) can exceed 10°C, which makes for a substantial degree of superheated steam. Hence

$$h_{V,1} = h_{gV_{\langle T_{B,1}, P_1 \rangle}} \tag{4.6.4}$$

where h_{gV} can be found from the superheated steam tables or relevant software such as NIST REFPROP [6].

$h_{B,1}$ is the enthalpy of the concentrated outlet at its temperature and salinity.

$$h_{B,1} = h_{fB_{\langle T_{B,1}, X_{B,1} \rangle}} \tag{4.6.5}$$

After rearranging Eq. (4.6) we have

$$\dot{m}_{\text{HS},1} \cdot \left(h_{\text{HS},1,\text{in}} - h_{\text{HS},1,\text{out}}\right) = \dot{m}_{V,1} \cdot h_{V,1} + \dot{m}_{B,1} \cdot h_{B,1} - \dot{m}_{F,1} \cdot h_{F,1} \tag{4.7}$$

The left-hand side of Eq. (4.7) is the amount of released energy from the heat source in the first effect $\left(\dot{Q}_{\text{HS},1}\right)$. With the combination of Eqs. (4.4) and (4.7) the energy balance equation for the first effect is written as

$$\dot{Q}_{\text{HS},1} = \dot{m}_{F,1} \cdot \left(\frac{1}{R} \cdot h_{V,1} + \left(\frac{R-1}{R} \right) \cdot h_{B,1} - h_{F,1} \right) \tag{4.8}$$

4.2.1.1.2 Temperature—Energy Profile and *UA* Value

Fig. 4.4 represents a general temperature—energy profile of the heat source and the feed across the first effect. In this effect, the heat source temperature drops across the evaporator, while the feed temperature first (preheating zone shown as zone 1) increases from the inlet temperature to the relevant boiling temperature ($T_{B^*,1} = T_{\text{sat}\langle P_1 \rangle} + BPE_{B^*,1}$), according to its concentration, at the prevailing pressure (P_1) (a sensible—sensible heat transfer) and then increases further during the evaporation process (evaporation zone, shown as zone 2) because of the continuing concentration that brings about the heightened boiling point elevation (BPE), to the relevant boiling temperature ($T_{B,1} = T_{\text{sat}\langle P_1 \rangle} + BPE_{B,1}$). Therefore due to the different types of heat transfer at the preheating and evaporation zones, there are two different UA values. In seawater

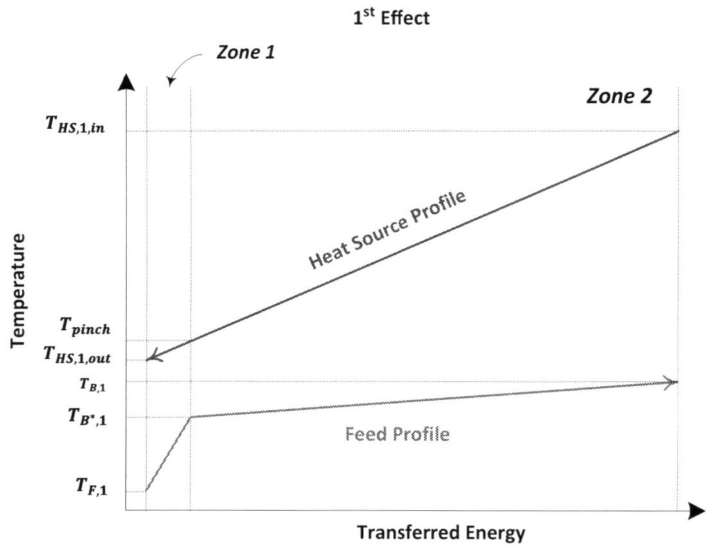

Figure 4.4
Temperature–energy profile for the first effect.

applications, due to small changes in the *BPE* during the evaporation process $\left(BPE_{B^*,1} \cong BPE_{B,1}\right)$ a conservative and reasonable approximation can be considered, namely that the feed temperature remains constant $\left(T_{B^*,1} = T_{B,1}\right)$. In these equations, $BPE_{B^*,1}$ and $BPE_{B,1}$ are the *BPEs* of the inlet feed and the outlet concentrated stream, respectively.

In our cases, namely, seawater desalination (Chapter 8) and process liquor reconcentration (Chapter 9), the preheating zone (zone 1) covers less than 5% of the total energy transferred and therefore to a good approximation it can be ignored. As a rough estimation, it can be seen that the enthalpy change for a unit mass of pure water in liquid phase for a 30°C increase in temperature from 30 to 60°C is equal to 125.45 kJ/kg, whereas the energy needed for the evaporation of a unit mass of pure water at 60°C is 2357.70 kJ/kg, which is around 19 times more than the sensible enthalpy (namely 125.45 kJ/kg) change. Hence the overall U value (heat transfer coefficient, $(\text{kW/m}^2\cdot\text{K})$) for the first effect is practically U_{z2}, which relates to the evaporation zone (zone 2), where sensible–latent heat transfer happens.

To calculate the *UA* values for the two different zones, the pinch point temperature of the first effect's evaporator (as shown in Fig. 4.4) should be calculated. For this purpose the energy balance for the first zone (Fig. 4.4) is written as

$$\dot{m}_{\text{HS},1} \cdot \left(h_{f_{\text{HS}\left\langle T_{\text{pinch}}, X_{\text{HS},1} \right\rangle}} - h_{f_{\text{HS}\left\langle T_{\text{HS},1,\text{out}}, X_{\text{HS},1} \right\rangle}} \right) = \dot{m}_{F,1} \cdot \left(h_{f_{F\left\langle T_{B^*,1}, X_{F,1} \right\rangle}} - h_{f_{F\left\langle T_{F,1}, X_{F,1} \right\rangle}} \right) \tag{4.9}$$

The right-hand-side term is the amount of energy received by the feed in the first zone, $\left(\dot{Q}_{F,1,z1}\right)$, and the left-hand side is the amount of released energy from the heat source in the same zone, $\left(\dot{Q}_{\mathrm{HS},1,z1}\right)$. Then, T_{pinch} is interpolated from Eq. (4.10):

$$h_{f_{\mathrm{HS}}\left\langle T_{\mathrm{pinch}},X_{\mathrm{HS},1}\right\rangle} = \frac{\dot{m}_{F,1}}{\dot{m}_{\mathrm{HS},1}} \cdot \left(h_{f_F\left\langle T_{B^*,1},X_{F,1}\right\rangle} - h_{f_F\left\langle T_{F,1},X_{F,1}\right\rangle} \right) + h_{f_{\mathrm{HS}}\left\langle T_{\mathrm{HS},1,\mathrm{out}},X_{\mathrm{HS},1}\right\rangle} \tag{4.10}$$

and the *UA* value can be calculated from

$$(UA)_{1,z1} = \frac{\dot{Q}_{\mathrm{HS},1,z1}}{\Delta T_{lm,1,z1}} \tag{4.11}$$

where

$$\Delta T_{lm,1,z1} = \frac{\left(T_{\mathrm{HS},1,\mathrm{out}} - T_{F,1}\right) - \left(T_{\mathrm{pinch}} - T_{B^*,1}\right)}{\ln\left(\dfrac{T_{\mathrm{HS},1,\mathrm{out}} - T_{F,1}}{T_{\mathrm{pinch}} - T_{B^*,1}}\right)} \tag{4.12}$$

The same methodology can be applied to the second zone to find out the relevant *UA* value, so that

$$(UA)_{1,z2} = \frac{\dot{Q}_{\mathrm{HS},1,z2}}{\Delta T_{lm,1,z2}} \tag{4.13}$$

where

$$\dot{Q}_{\mathrm{HS},1,z2} = \dot{m}_{\mathrm{HS},1} \cdot \left(h_{f_{\mathrm{HS}}\left\langle T_{\mathrm{HS},1,\mathrm{in}},X_{\mathrm{HS},1}\right\rangle} - h_{f_{\mathrm{HS}}\left\langle T_{\mathrm{pinch}},X_{\mathrm{HS},1}\right\rangle} \right) \tag{4.14}$$

$$\Delta T_{lm,1,z2} = \frac{\left(T_{\mathrm{HS},1,\mathrm{in}} - T_{B,1}\right) - \left(T_{\mathrm{pinch}} - T_{B^*,1}\right)}{\ln\left(\dfrac{T_{\mathrm{HS},1,\mathrm{in}} - T_{B,1}}{T_{\mathrm{pinch}} - T_{B^*,1}}\right)} \tag{4.15}$$

As mentioned before, $(UA)_{1,z2} \gg (UA)_{1,z1}$, which means to a good approximation

$$(UA)_1 \approx (UA)_{1,z2} \tag{4.16}$$

4.2.1.2 Second Effect to the Last Effect

Fig. 4.5 represents the inlets and outlets of the second to the last effects. As mentioned earlier, the produced vapor in each effect is used as the heat source for the following effect. The vapor inside the tubes condenses and its energy is released to evaporate the inlet feed water, which is distributed from the top onto the hot pipes or plates (depending

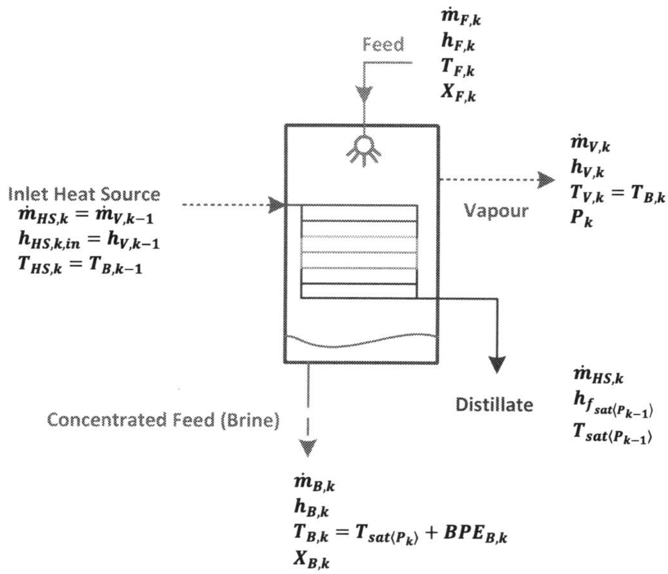

Figure 4.5
Schematic design of the second to the last effects ($k \in \{2,\ldots,n\}$).

on the type of evaporator). Depending on the degree of superheating, the superheated vapor desuperheats to the relevant saturation temperature at its prevailing pressure $\left(T_{\text{sat}\langle P_k\rangle}\right)$ and then it condenses (at a constant pressure). As with the first effect, the produced vapor goes on to power the following effect with the resulting concentrated feed being drained from the bottom of that effect.

4.2.1.2.1 Energy Balance

As shown in Fig. 4.5, the energy balance for these effects can be expressed as

$$\dot{m}_{\text{HS},k} \cdot \left(h_{\text{HS},k,\text{in}} - h_{f_{\text{sat}\langle P_{k-1}\rangle}} \right) = \dot{m}_{V,k} \cdot h_{V,k} + \dot{m}_{B,k} \cdot h_{B,k} - \dot{m}_{F,k} \cdot h_{F,k} \tag{4.17}$$

The left-hand side of Eq. (4.17) is the amount of released energy from the heat source $\left(\dot{Q}_{\text{HS},k}\right)$. As can be seen, the difference between Eq. (4.17) and Eq. (4.7) is only related to the left-hand side, which pertains to the different type of heat transfer. In the first effect the heat source temperature drops while its phase does not change, but for the second to the last effects the vaporized heat source is first desuperheated and then condensed. After rearranging and combining with Eq. (4.4), it can be written as

$$\dot{Q}_{\text{HS},k} = \dot{m}_{F,k} \left(\frac{1}{R} \cdot h_{V,k} + \left(\frac{R-1}{R} \right) h_{B,k} - h_{F,k} \right) \tag{4.18}$$

4.2.1.2.2 Temperature–Energy Profile and *UA* Value

Fig. 4.6 shows a general temperature–energy profile across the second effect to the last effect. As explained before, the heat source of these effects is the superheated vapor from the previous effect. Referring to this figure, the superheated vapor is first desuperheated to the relevant saturated temperature of the previous effect in zone 1 (desuperheating zone), and then condenses. In contrast, the trend of feed temperature profile is the same as in the first effect if the saturation pressure of the feed stream is less than the effect's pressure, as shown in Fig. 4.6A. Therefore, in this case, the feed temperature increases to the relevant boiling temperature in the preheating zone (zones 1 and 2) and then its temperature increases further during the evaporation process (zone 3) (however, as mentioned in Section 4.2.1.1.2, with a good approximation it can be considered to remain constant in seawater applications). In zone 1, during the desuperheating process, the feed water temperature rises up to $T_{F^*,k}$. In both seawater and process liquor applications, the feed water temperature cannot reach its boiling point temperature in this zone, therefore $T_{F^*,k} < T_{B^*,k}$. The area, however, that is needed for desuperheating has a significant effect on the capital cost of the evaporator; this will be discussed in detail in Chapter 9 for process liquor reconcentration applications, wherein the degree of superheat is much higher than that in seawater applications. In these effects, as mentioned before, a reasonable approximation for the overall U value is simply that of the evaporation zone (zone 3).

As mentioned earlier, in zone 1 of Fig. 4.6A, the feed water temperature will reach $T_{F^*,k}$, which is lower than its boiling point temperature. $T_{F^*,k}$ can be interpolated from the energy balance equation, which is written as

$$h_{f_F\left\langle T_{F^*,k},X_{F,k}\right\rangle} = \frac{\dot{m}_{\mathrm{HS},k}}{\dot{m}_{F,k}} \cdot \left(h_{\mathrm{HS},k,\mathrm{in}} - h_{g_{\mathrm{sat}\left\langle P_{k-1}\right\rangle}}\right) + h_{f_F\left\langle T_{F,k},X_{F,k}\right\rangle} \tag{4.19}$$

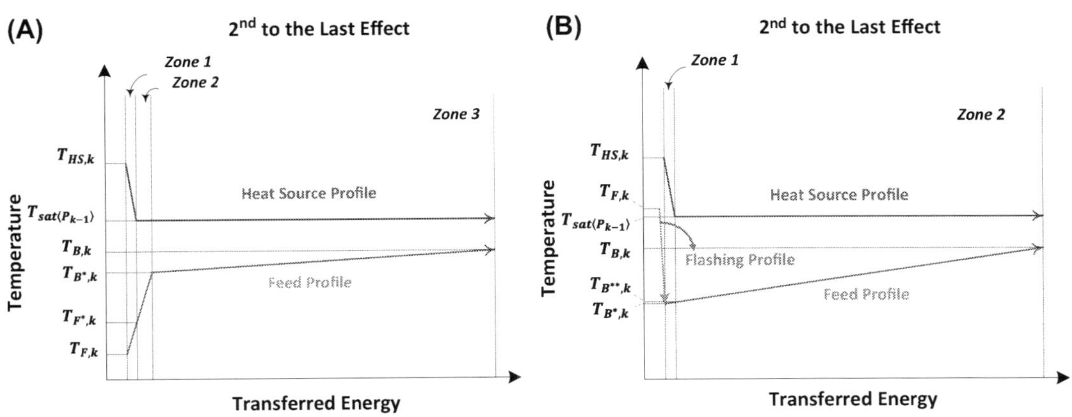

Figure 4.6

(A)Temperature–energy profile for the second to the last effect without flashed feed stream. (B) Temperature–energy profile for the second to the last effect with flashed feed stream.

where:

$$h_{\mathrm{HS},k,\mathrm{in}} = h_{gv_{\langle T_{\mathrm{HS},k}, P_{k-1}\rangle}} \tag{4.19.1}$$

Therefore, the *UA* value can be calculated from

$$(UA)_{k,z1} = \frac{\dot{m}_{\mathrm{HS},k} \cdot \left(h_{\mathrm{HS},k,\mathrm{in}} - h_{g_{\mathrm{sat}\langle P_{k-1}\rangle}} \right)}{\Delta T_{lm,k,z1}} \tag{4.20}$$

where

$$\Delta T_{lm,k,z1} = \frac{\left(T_{\mathrm{sat}\langle P_{k-1}\rangle} - T_{F^*,k} \right) - \left(T_{\mathrm{HS},k} - T_{F,k} \right)}{\ln\left(\dfrac{T_{\mathrm{sat}\langle P_{k-1}\rangle} - T_{F^*,k}}{T_{\mathrm{HS},k} - T_{F,k}} \right)} \tag{4.21}$$

In zone 2, the feed water temperature rises to its relevant boiling point while the vapor inside the tube is condensed. Therefore the *UA* value can be found from

$$(UA)_{k,z2} = \frac{\dot{m}_{F,k} \cdot \left(h_{f_{F\langle T_{B^*,k}, X_{F,k}\rangle}} - h_{f_{F\langle T_{F^*,k}, X_{F,k}\rangle}} \right)}{\Delta T_{lm,k,z2}} \tag{4.22}$$

where

$$\Delta T_{lm,k,z2} = \frac{\left(T_{\mathrm{sat}\langle P_{k-1}\rangle} - T_{B^*,k} \right) - \left(T_{\mathrm{sat}\langle P_{k-1}\rangle} - T_{F^*,k} \right)}{\ln\left(\dfrac{T_{\mathrm{sat}\langle P_{k-1}\rangle} - T_{B^*,k}}{T_{\mathrm{sat}\langle P_{k-1}\rangle} - T_{F^*,k}} \right)} \tag{4.23}$$

The quality, x, of the vapor inside the tube at the threshold between the second and the third zone also can be calculated from the enthalpy of vapor at that point, according to the energy balance equation, which is written as

$$h'_{g_k} = h_{g_{\mathrm{sat}\langle P_{k-1}\rangle}} - \frac{\dot{m}_{F,k}}{\dot{m}_{\mathrm{HS},k}} \cdot \left(h_{f_{F\langle T_{B^*,k}, X_{F,k}\rangle}} - h_{f_{F\langle T_{F^*,k}, X_{F,k}\rangle}} \right) \tag{4.24}$$

Therefore,

$$x = \frac{h'_{g_k} - h_{f_{\mathrm{sat}\langle P_{k-1}\rangle}}}{h_{g_{\mathrm{sat}\langle P_{k-1}\rangle}} - h_{f_{\mathrm{sat}\langle P_{k-1}\rangle}}} \tag{4.25}$$

which in our applications the quality, x, is not less than 98%.

Finally, for zone 3, the relevant *UA* value can be calculated as

$$(UA)_{k,z3} = \frac{\dot{m}_{\mathrm{HS},k} \cdot \left(h'_{g_k} - h_{f_{\mathrm{sat}\langle P_{k-1}\rangle}}\right)}{\Delta T_{k,z3}} \tag{4.26}$$

where

$$\Delta T_{lm,k,z3} = \frac{\left(T_{\mathrm{sat}\langle P_{k-1}\rangle} - T_{B^*,k}\right) - \left(T_{\mathrm{sat}\langle P_{k-1}\rangle} - T_{B,k}\right)}{\ln\left(\dfrac{T_{\mathrm{sat}\langle P_{k-1}\rangle} - T_{B^*,k}}{T_{\mathrm{sat}\langle P_{k-1}\rangle} - T_{B,k}}\right)} \tag{4.27}$$

In seawater applications since the evaporation occurs at effectively a constant temperature, a constant temperature difference between the heat source and feed water $\left(\Delta T_{lm,k,z3} = T_{\mathrm{sat}\langle P_{k-1}\rangle} - T_{B,k}\right)$ exists.

Sometimes the feed saturation pressure is more than the effect's pressure, as per what is shown in Fig. 4.6B. In this case, before the feed is being distributed onto the tubes, a negligible amount of feed (less than 1% in most cases; see Chapter 9) is flashed while its pressure and temperature decrease. For these effects the inlet feed temperature is considered to equate to the relevant feed boiling temperature, $T_{B^*,k}$ (which is related to its concentration), which corresponds to the pressure in the relevant effects, shown in Fig. 4.6B. After the temperature reduction, once it distributes onto the tubes, evaporation occurs.

As shown in Fig. 4.6B in zone 1 of these effects, the amount of energy required for desuperheating of the heat source vapor is negligible (see Chapter 9). During the desuperheating process, the feed stream temperature rises to $T_{B^{**},k}$. In our case, due to the small amount of energy released in the desuperheating zone, a reasonable approximation can be considered, namely $T_{B^*,k} = T_{B^{**},k}$.

When feed stream flashes at the entrance of the effect as shown in Fig. 4.6B, the *UA* value for zone 1 can be calculated from

$$(UA)_{k,z1} = \frac{\dot{m}_{\mathrm{HS},k} \cdot \left(h_{gv_{\langle T_{\mathrm{HS},k}, P_{k-1}\rangle}} - h_{g_{\mathrm{sat}\langle P_{k-1}\rangle}}\right)}{\Delta T_{lm,k,z1}} \tag{4.28}$$

where

$$\Delta T_{lm,k,z1} = \frac{\left(T_{\mathrm{HS},k} - T_{B^*,k}\right) - \left(T_{\mathrm{sat}\langle P_{k-1}\rangle} - T_{B^*,k}\right)}{\ln\left(\dfrac{T_{\mathrm{HS},k} - T_{B^*,k}}{T_{\mathrm{sat}\langle P_{k-1}\rangle} - T_{B^*,k}}\right)} \tag{4.29}$$

In zone 2, we have

$$(UA)_{k,z2} = \frac{\dot{m}_{\mathrm{HS},k} \cdot \left(h_{g_{\mathrm{sat}\langle P_{k-1}\rangle}} - h_{f_{\mathrm{sat}\langle P_{k-1}\rangle}}\right)}{\Delta T_{lm,k,z2}} \tag{4.30}$$

where

$$\Delta T_{lm,k,z2} = \frac{\left(T_{\mathrm{sat}\langle P_{k-1}\rangle} - T_{B^*,k}\right) - \left(T_{\mathrm{sat}\langle P_{k-1}\rangle} - T_{B,k}\right)}{\ln\left(\frac{T_{\mathrm{sat}\langle P_{k-1}\rangle} - T_{B^*,k}}{T_{\mathrm{sat}\langle P_{k-1}\rangle} - T_{B,k}}\right)} \tag{4.31}$$

4.2.1.3 Condenser

The condenser (Fig. 4.7) is located after the last effect to condense the last remaining produced vapor (from the last effect). In seawater application the coolant is seawater. In our case, for all the seawater application simulations, the inlet seawater temperature has been considered to be 28°C. The cooling water absorbs the released energy from the condensation of the superheated vapor inside the heat exchanger. A portion of the outlet cooling water, which has been preheated in the condenser, is used then as the feed water for the desalination process (in seawater application) and the balance is purged to the sea. In mineral refinery applications (Chapter 9) the outlet cooling water is rejected to the evaporation pond, otherwise known as the recovery lake.

4.2.1.3.1 Energy Balance

The amount of required coolant mass flow rate $(\dot{m}_C)$ is calculated from the energy balance equation, which is written as:

$$\dot{m}_{\mathrm{HS,cond}} \cdot \left(h_{\mathrm{HS,cond,in}} - h_{\mathrm{HS,cond,out}}\right) = \dot{m}_C \cdot \left(h_{C,\mathrm{out}} - h_{C,\mathrm{in}}\right) \tag{4.32}$$

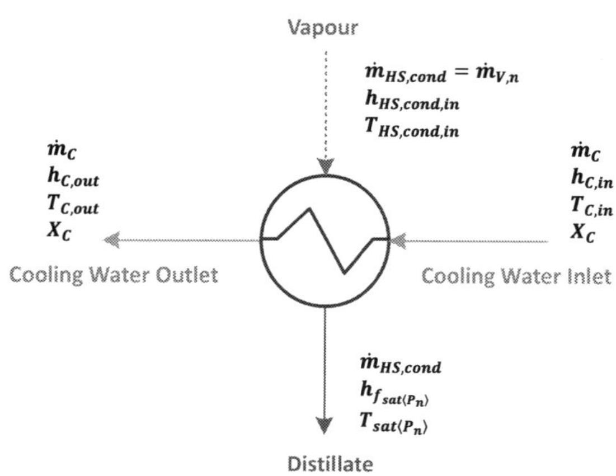

Figure 4.7
Schematic design of the condenser.

where the left-hand-side term is the amount of energy released from the hot side $\left(\dot{Q}_{\mathrm{HS,cond}}\right)$ and

$$\dot{m}_{\mathrm{HS,cond}} = \dot{m}_{V,n} \tag{4.32.1}$$

$$h_{\mathrm{HS,cond,in}} = h_{V,n} \tag{4.32.2}$$

$$h_{\mathrm{HS,cond,out}} = h_{f_{\mathrm{sat}\langle P_n \rangle}} \tag{4.32.3}$$

$h_{C,\mathrm{in}}$ and $h_{C,\mathrm{out}}$ are the enthalpies of the inlet and outlet cooling water, which are a function of their temperatures and salinities, respectively.

$$h_{C,\mathrm{in}} = h_{f_{C\langle T_{C,\mathrm{in}}, X_C \rangle}} \tag{4.32.4}$$

$$h_{C,\mathrm{out}} = h_{f_{C\langle T_{C,\mathrm{out}}, X_C \rangle}} \tag{4.32.5}$$

4.2.1.3.2 Temperature—Energy Profile and *UA* Value

As shown in Fig. 4.8 the cooling water temperature increases across the condenser while the trend of the inlet superheated vapor temperature profile is the same as the other effects, so that zone 1 (namely the desuperheating zone) is negligible once again, with the overall heat transfer coefficient being practically U_2, where latent-sensible heat transfer happens. The impact of desuperheating is being considered as the extra required heat exchanger area for the mineral refinery applications (see Chapter 9).

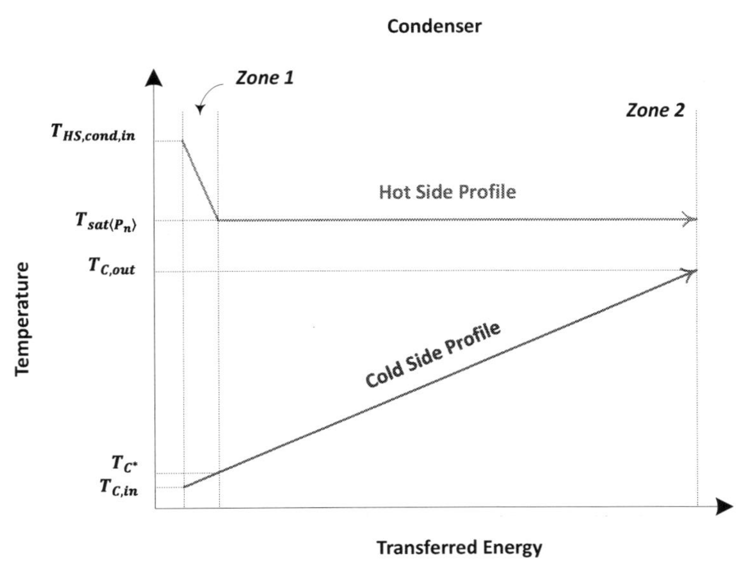

Figure 4.8
Temperature—energy profile for the condenser.

In zone 1 (Fig. 4.8) the cooling water temperature will reach T_{C^*}, which can be found from the energy balance equation, which for this zone is

$$\dot{m}_{\text{HS,cond}} \cdot \left(h_{\text{HS,cond,in}} - h_{g_{\text{sat}\langle P_n \rangle}} \right) = \dot{m}_C \cdot \left(h_{C^*} - h_{C,\text{in}} \right) \tag{4.33}$$

Rearranging, we have

$$h_{C^*} = \frac{\dot{m}_{\text{HS,cond}}}{\dot{m}_C} \cdot \left(h_{\text{HS,cond,in}} - h_{g_{\text{sat}\langle P_n \rangle}} \right) + h_{C,\text{in}} \tag{4.34}$$

h_{C^*} can be written as

$$h_{C^*} = h_{f_{C\langle T_{C^*}, X_C \rangle}} \tag{4.35}$$

T_{C^*} can then be calculated from Eq. (4.35).

With the same methodology as the other effects, the *UA* value for zone 1 is calculated as

$$(UA)_{k,z1} = \frac{\dot{m}_{\text{HS,cond}} \cdot \left(h_{\text{HS,cond,in}} - h_{g_{\text{sat}\langle P_n \rangle}} \right)}{\Delta T_{lm,\text{cond},z1}} \tag{4.36}$$

where

$$\Delta T_{lm,\text{cond},z1} = \frac{\left(T_{\text{sat}\langle P_n \rangle} - T_{C^*} \right) - \left(T_{\text{HS,cond,in}} - T_{C,\text{in}} \right)}{\ln \left(\frac{T_{\text{sat}\langle P_n \rangle} - T_{C^*}}{T_{\text{HS,cond,in}} - T_{C,\text{in}}} \right)} \tag{4.37}$$

In zone 2, the cooling water (cold side medium) temperature raises up to the outlet temperature while the vapor (hot side medium) is condensed. Therefore the relevant *UA* value is calculated as

$$(UA)_{k,z2} = \frac{\dot{m}_{\text{HS,cond}} \cdot \left(h_{g_{\text{sat}\langle P_n \rangle}} - h_{f_{\text{sat}\langle P_n \rangle}} \right)}{\Delta T_{lm,\text{cond},z2}} \tag{4.38}$$

where

$$\Delta T_{lm,\text{cond},z2} = \frac{\left(T_{\text{sat}\langle P_n \rangle} - T_{C^*} \right) - \left(T_{\text{sat}\langle P_n \rangle} - T_{C,\text{out}} \right)}{\ln \left(\frac{T_{\text{sat}\langle P_n \rangle} - T_{C^*}}{T_{\text{sat}\langle P_n \rangle} - T_{C,\text{out}}} \right)} \tag{4.39}$$

4.2.2 Preheated Multi-Effect Distillation Process

As shown in Fig. 4.9, the preheated MED includes a primary MED section and some liquid–liquid heat exchangers. The mass and energy balance for the primary MED section

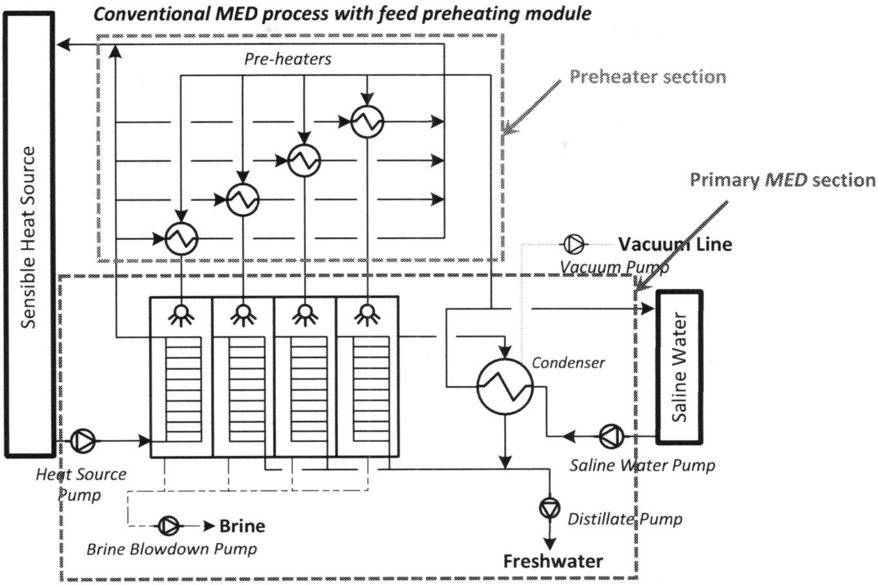

Figure 4.9
Sections of a preheated multi-effect distillation system.

including the effects and the condenser are the same as detailed in Sections 4.2.1.1, 4.2.1.2, and 4.2.1.3.

The preheaters, which can be plate-type heat exchangers, harvest a portion of the remaining energy that is not extracted from the heat source at the first effect of the primary MED section, and delivers it to the feed streams. A typical 3°C approach temperature has been considered in all simulations for these types of heat exchanger.

As shown in Fig. 4.10 the hot side outlet temperature ($T_{\text{h\_prh},i,\text{out}}$) is calculated from the energy balance Eq. (4.40).

$$\dot{m}_{\text{h\_prh},i} \cdot \left(h_{\text{h\_prh},i,\text{in}} - h_{\text{h\_prh},i,\text{out}}\right) = \dot{m}_{\text{c\_prh},i} \cdot \left(h_{\text{c\_prh},i,\text{out}} - h_{\text{c\_prh},i,\text{in}}\right) \tag{4.40}$$

where i is the preheater number, h_prh and c_prh mean the hot side and the cold side of the preheater, respectively, where

$$h_{\text{h\_prh},i,\text{in}} = h_{\text{HS},1,\text{out}} = h_{f_{\text{HS}}\left\langle T_{\text{HS},1,\text{out}}, X_{\text{HS},1}\right\rangle} \tag{4.40.1}$$

$$h_{\text{h\_prh},i,\text{out}} = h_{f_{\text{HS}}\left\langle T_{\text{h\_prh},i,\text{out}}, X_{\text{HS},1}\right\rangle} \tag{4.40.2}$$

By knowing the hot side outlet enthalpy from Eq. (4.40), $T_{\text{h\_prh},i,\text{out}}$ can be found from Eq. (4.40.2). These equations can be written for each preheater and the relevant hot side outlet temperatures can then be calculated.

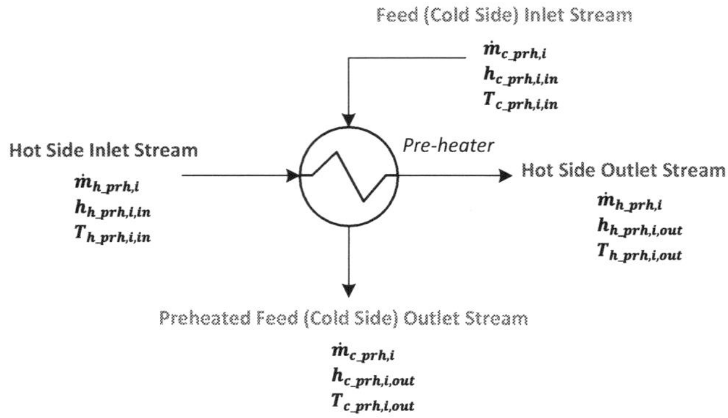

Figure 4.10
Schematic design of a preheater.

4.2.3 Boosted Multi-Effect Distillation Process

As shown in Fig. 4.11, the B-MED includes a primary MED section (including an injected effect to which the vapor produced by the booster is injected) and a booster that is the same as the first effect of the primary section.

4.2.3.1 Primary Multi-Effect Distillation Section

As shown in Fig. 4.11, this section includes three subsections: the noninjected effects, the injected effect, and the condenser. The mass and energy balance, temperature—energy

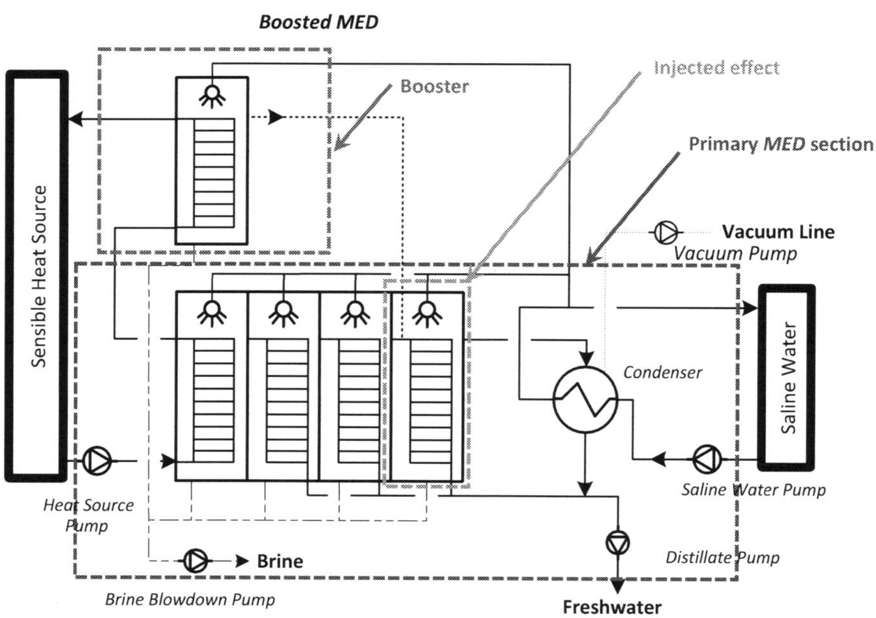

Figure 4.11
Sections of a boosted multi-effect distillation system.

profiles, and *UA* values for the noninjected effects and the condenser are the same as detailed in Sections 4.2.1.1, 4.2.1.2, and 4.2.1.3.

For the injected effect, the amount of injected vapor (vapor generated at the booster) is added to the produced vapor of its previous effect, and together they will be used as the heat source of the injected effect (Fig. 4.12). The driving force for injection is considered to be 500 Pa pressure difference to make for a natural vapor flow and compensate for the relevant pressure drops [3,4]. The temperature of the mixed vapor streams is calculated based on an energy balance equation. It is assumed that the pressure of the mixed streams is equal to the lowest pressure, which means the pressure of the previous effect (P_{k-1} in Fig. 4.12); therefore just as with the noninjected effects, a superheated vapor is available but with a slightly higher degree of superheating. With this assumption, for the effect with vapor injection, the condensation occurs at the previous effect's pressure, P_{k-1} (the same as the noninjected effects).

4.2.3.1.1 Energy Balance for the Injected Effect

As it shows in Fig. 4.12, the energy balance for the injected effect can be figured out as

$$\dot{m}_{\mathrm{HS},k} \cdot \left(h_{\mathrm{HS},k,in} - h_{f_{\mathrm{sat}\langle P_{k-1}\rangle}}\right) = \dot{m}_{V,k} \cdot h_{V,k} + \dot{m}_{B,k} \cdot h_{B,k} - \dot{m}_{F,k} \cdot h_{F,k} \tag{4.41}$$

The left-hand side of Eq. (4.41) is the amount of released energy from the heat source $\left(\dot{Q}_{\mathrm{HS},k}\right)$. After rearranging and combining with Eq. (4.4) it can be written as

$$\dot{Q}_{\mathrm{HS},k} = \dot{m}_{F,k}\left(\frac{1}{R} \cdot h_{V,k} + \left(\frac{R-1}{R}\right) h_{B,k} - h_{F,k}\right) \tag{4.42}$$

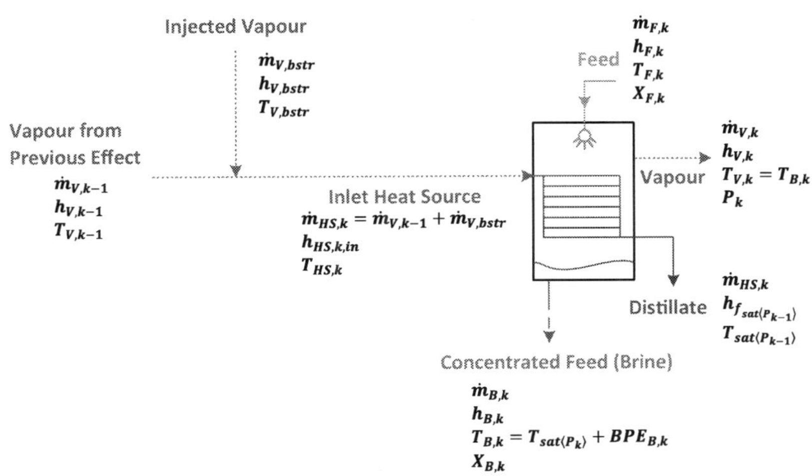

Figure 4.12

Schematic design of the injected effect in a boosted multi-effect distillation process ($k \in \{2,\ldots,n\}$).

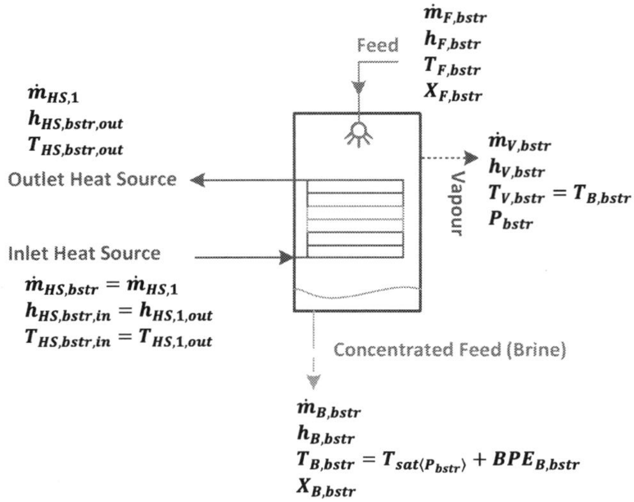

Figure 4.13
Schematic design of the booster.

For the injected effect the heat source inlet mass flow rate is calculated as

$$\dot{m}_{\mathrm{HS},k} = \dot{m}_{\mathrm{V},k-1} + \dot{m}_{\mathrm{V,bstr}} \tag{4.43}$$

The heat source inlet energy for the injected effects can be also written as

$$h_{\mathrm{HS},k,\mathrm{in}} = \frac{\dot{m}_{\mathrm{V},k-1} \cdot h_{\mathrm{V},k-1} + \dot{m}_{\mathrm{V,bstr}} \cdot h_{\mathrm{V,bstr}}}{\dot{m}_{\mathrm{V},k-1} + \dot{m}_{\mathrm{V,bstr}}} \tag{4.44}$$

where $h_{\mathrm{V,bstr}}$ is the enthalpy of the produced vapor in the booster effect.

4.2.3.1.2 Temperature–Energy Profile and *UA* Value for the Injected Effect

The temperature–energy profile (Fig. 4.6) and all the equations that have been explained in Section 4.2.1.2.2 are used for the analysis of the injected effect. The only issue is that the impact of the injected vapor from the booster that mixes with the superheated vapor from the relevant upstream MED effect has to be considered in arriving at the mixed superheated vapor temperature. However, this impact is negligible for seawater application since the *BPE* is very low.

4.2.3.2 Booster

The same equations that have been explained in Section 4.2.1.1 for the first effect are used for the booster but with the parameters shown in Fig. 4.13.

4.2.4 Flash-Boosted Multi-Effect Distillation Process

As shown in Fig. 4.14, the FB-MED process can be divided into two main sections, the primary MED section (including the injected effects, noninjected effects, and the

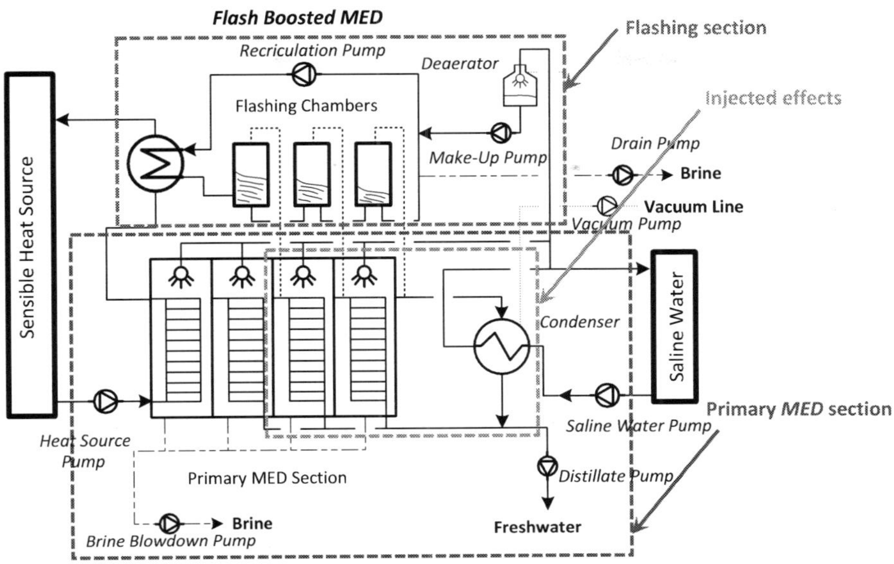

Figure 4.14
Schematic design of the flash-boosted multi-effect distillation process.

condenser) and the flashing section (including a battery of flashing chambers, brine recirculation cycle, brine heater, and deaerator).

4.2.4.1 Primary Multi-Effect Distillation Section

This section includes three subsections: the injected effects, noninjected effects, and the condenser. The mass and energy balance, temperature–energy profiles, and *UA* values for the noninjected effects are the same as detailed in Sections 4.2.1.1 and 4.2.1.2.

4.2.4.1.1 Injected Effects

For those effects with vapor injection from the flashing chambers, the injected vapor is added to the produced vapor of the previous effect, which will then be used as a heat source of that effect (see Fig. 4.15). The driving force for injection is considered to be 500 Pa pressure difference to make for a natural vapor flow and compensate for the relevant pressure drops [3,4]. The temperature of the mixed vapor streams is calculated based on an energy balance equation. It is assumed that the pressure of the mixed stream is equal to the lowest pressure, which means the pressure of the previous effect; therefore just as compared to the noninjected effects, a superheated vapor is available but with a slightly higher degree of superheating, which means, for the injected effect, the condensation occurs at the previous effect's pressure, P_{k-1}.

4.2.4.1.1.1 Energy Balance for the Injected Effects As shown in Fig. 4.15, the energy balance for these effects can be written as

$$\dot{m}_{\mathrm{HS},k}\cdot\left(h_{\mathrm{HS},k,\mathrm{in}}-h_{f_{\mathrm{sat}\left(P_{k-1}\right)}}\right)=\dot{m}_{V,k}\cdot h_{V,k}+\dot{m}_{B,k}\cdot h_{B,k}-\dot{m}_{F,k}\cdot h_{F,k} \tag{4.45}$$

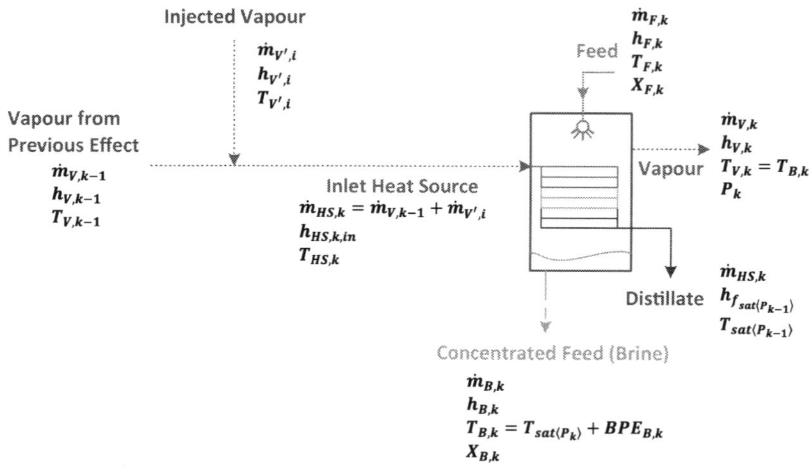

Figure 4.15
Schematic design of the injected effect in a flash-boosted multi-effect distillation process ($k \in \{2,\ldots,n\}$; $i \in \{1,\ldots,j\}$).

The left-hand side of Eq. (4.45) is the amount of released energy from the heat source $\left(\dot{Q}_{\mathrm{HS},k}\right)$. After rearranging and combining with Eq. (4.4) it can be written as

$$\dot{Q}_{\mathrm{HS},k} = \dot{m}_{F,k}\left(\frac{1}{R}\cdot h_{V,k} + \left(\frac{R-1}{R}\right)h_{B,k} - h_{F,k}\right) \tag{4.46}$$

As shown in Fig. 4.15, for the injected effects the heat source inlet mass flow rate is calculated by ($i \in \{1,\ldots,j\}$):

$$\dot{m}_{\mathrm{HS},k} = \dot{m}_{V,k-1} + \dot{m}_{V',i} \tag{4.47}$$

where i is the relevant flashing chamber's number where in an optimized seawater application it can be equivalent to $k-2$, depending on the design parameters, heat source temperature drop, and feed water properties.

The heat source inlet energy for the injected effects can also be written as

$$h_{\mathrm{HS},k,\mathrm{in}} = \frac{\dot{m}_{V,k-1}\cdot h_{V,k-1} + \dot{m}_{V',i}\cdot h_{V',i}}{\dot{m}_{V,k-1} + \dot{m}_{V',i}} \tag{4.48}$$

where $h_{V',i}$ is the enthalpy of the produced flashed vapor in the flashing chambers as explained in Section 4.2.4.2.

4.2.4.1.1.2 Temperature—Energy Profile and UA *Value for the Injected Effects* The temperature—energy profile (see Fig. 4.6) and all the equations that were explained in Section 4.2.1.2.2 are used for the injected effects of the FB-MED process. For these effects, the impact of the injected vapor from the flash vessels that mixes with the superheated vapor from the relevant upstream MED effect, has to be considered in arriving at the mixed superheated vapor temperature. This impact is negligible for seawater applications since the *BPE* is low. However it should be carefully considered in mineral refinery applications (see Chapter 9).

4.2.4.1.2 Condenser

In an FB-MED process the final mixture of injected vapor from the last stage of the flashing chamber (stage number j) and the produced vapor of the last effect (effect number n) will be channeled to the condenser (Fig. 4.16). In seawater applications the coolant is seawater. In our case for all seawater application simulations the inlet seawater temperature has been considered to be 28°C. The cooling water absorbs the released energy from the condensation of the superheated vapor inside the heat exchanger. A portion of the outlet cooling water that has been preheated in the condenser is then used as the feed water for the desalination process (in seawater applications) and the majority of balance outlet cooling water is returned to the sea except for a fraction, which will be used as the make-up water in the brine recirculation cycle (see Section 4.2.4.2).

4.2.4.1.2.1 Energy Balance The amount of required coolant mass flow rate ($\dot{m}_C$) is calculated from the energy balance equation, which is written as

$$\dot{m}_{\mathrm{HS,cond}} \cdot \left(h_{\mathrm{HS,cond,in}} - h_{\mathrm{HS,cond,out}}\right) = \dot{m}_C \cdot \left(h_{C,\mathrm{out}} - h_{C,\mathrm{in}}\right) \tag{4.49}$$

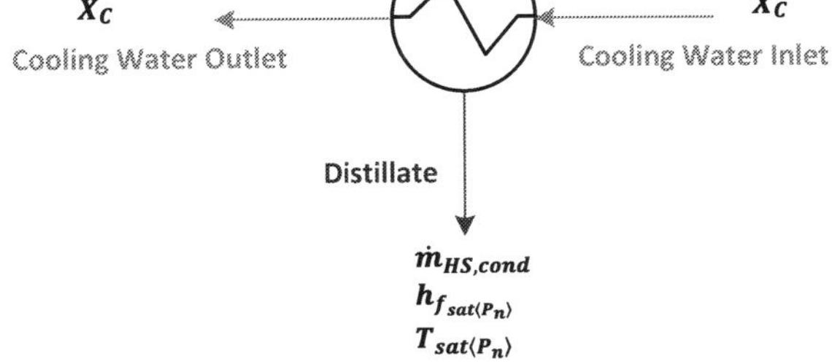

Figure 4.16
Schematic design of the condenser with vapor injection in a flash-boosted multi-effect distillation process.

where the left-hand-side term is the amount of energy released from the hot side $\left(\dot{Q}_{\text{HS,cond}}\right)$ and

$$\dot{m}_{\text{HS,cond}} = \dot{m}_{V,n} + \dot{m}_{V',j} \tag{4.49.1}$$

$$h_{\text{HS,cond,in}} = \frac{\dot{m}_{V,n} \cdot h_{V,n} + \dot{m}_{V',j} \cdot h_{V',j}}{\dot{m}_{V,n} + \dot{m}_{V',j}} \tag{4.49.2}$$

$$h_{\text{HS,cond,out}} = h_{f_{\text{sat}\langle P_n \rangle}} \tag{4.49.3}$$

$h_{C,\text{in}}$ and $h_{C,\text{out}}$ are the enthalpies of the inlet and outlet cooling water, which are a function of their temperatures and salinities, respectively.

$$h_{C,\text{in}} = h_{f_{C\langle T_{C,\text{in}}, X_C \rangle}} \tag{4.49.4}$$

$$h_{C,\text{out}} = h_{f_{C\langle T_{C,\text{out}}, X_C \rangle}} \tag{4.49.5}$$

4.2.4.1.2.2 Temperature—Energy Profile and UA *Value* The temperature—energy profile and *UA* value are the same as detailed in Section 4.2.1.3.2, however the impact of vapor injection should be considered in the inlet vapor mixture of the condenser.

4.2.4.2 Flashing Section

This section constitutes the core of the FB-MED system. It harvests the remaining available energy from the heat source (as compared to the conventional and B-MED processes), and distributes it to the primary MED effects with the aim of increasing the production rate. As mentioned before, the generated vapor from each stage of flashing is injected into the judicious effect of the primary MED plant for further boosting as in the previous steam-boosted scheme. It is discernible that this scheme permits the waste heat source outlet temperature to approach the mixing point temperature between the flashing chamber outlet and the condenser outlet temperatures as close as is practicable. This process therefore engenders a maximally efficient scheme to exploit the potential of waste heat, far more than the conventional and B-MED technologies [5].

The flashing section includes a liquid—liquid heat exchanger as the brine heater, a battery of flashing chambers, a brine recirculation process to decrease the make-up feed water flow rate, and a deaerator to reduce the dissolved gases in the make-up water stream (Fig. 4.17).

4.2.4.2.1 Brine Heater

This heat exchanger (the liquid—liquid heat exchanger in Fig. 4.18), which can be a plate-type heat exchanger, harvests the remaining available energy which is not extracted from the heat source by the first effect of the primary MED section, and delivers it to the flashing chambers. A typical 3°C temperature of approach has been considered in all the simulations for this heat exchanger.

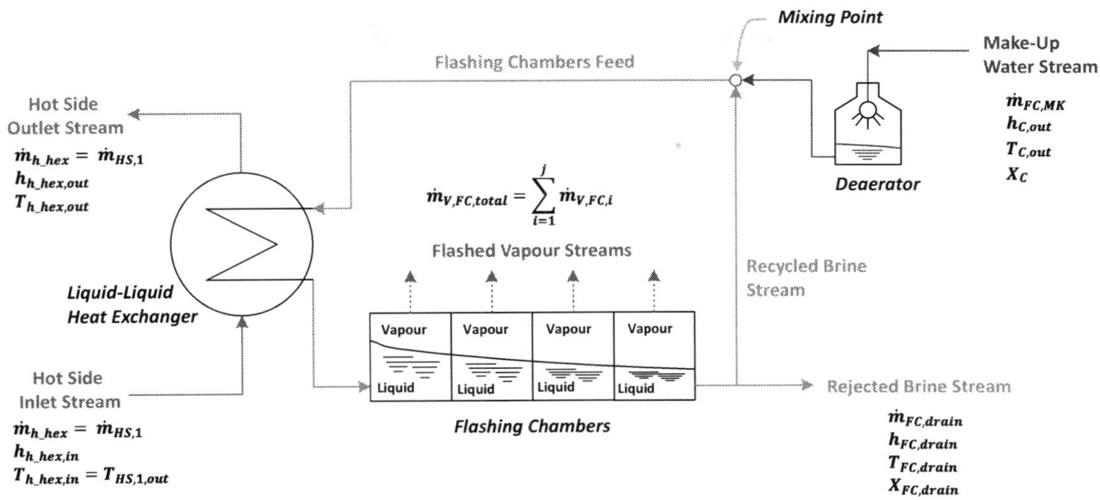

Figure 4.17
Schematic design of the flashing section.

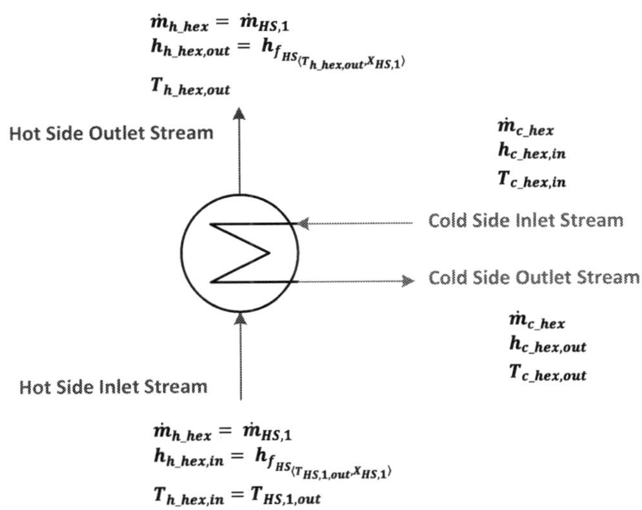

Figure 4.18
Schematic design of the brine heater in a flash-boosted multi-effect distillation process.

By knowing the heat source condition, the amount of required feed flow rate through the flashing chambers ($\dot{m}_{\text{c\_hex}}$) can be calculated from the energy balance of this heat exchanger and expressed as

$$\dot{m}_{\text{h\_hex}} \cdot \left(h_{\text{h\_hex,in}} - h_{\text{h\_hex,out}}\right) = \dot{m}_{\text{c\_hex}} \cdot \left(h_{\text{c\_hex,out}} - h_{\text{c\_hex,in}}\right) \tag{4.50}$$

where

$$\dot{m}_{h\_hex} = \dot{m}_{HS,1} \tag{4.50.1}$$

$$h_{h\_hex,in} = h_{HS,1,out} = h_{f_{HS}}\langle T_{HS,1,out}, X_{HS,1}\rangle \tag{4.50.2}$$

$$h_{h\_hex,out} = h_{f_{HS}}\langle T_{h\_hex,out}, X_{HS,1}\rangle \tag{4.50.3}$$

4.2.4.2.2 Battery of Flashing Chambers

It is well known that when a pressurized liquid enters a flashing chamber with a pressure lower than the liquid's saturated pressure, the liquid becomes superheated and vigorously flashes (Fig. 4.19). During the flashing phenomenon a portion of the liquid evaporates while the temperature and pressure decrease until the equilibrium condition is achieved.

The mass and salinity balance for a flashing chamber are written as

$$\dot{m}_{FC,i,in} = \dot{m}_{V,FC,i} + \dot{m}_{FC,i,out} \tag{4.51}$$

$$\dot{m}_{FC,i,in} \cdot X_{FC,i,in} = \dot{m}_{FC,i,out} \cdot X_{FC,i,out} \tag{4.52}$$

The energy balance is

$$\dot{m}_{FC,i,in} \cdot h_{FC,i,in} = \dot{m}_{FC,i,out} \cdot h_{FC,i,out} + \dot{m}_{V,FC,i} \cdot h_{V,FC,i} \tag{4.53}$$

where

$$h_{FC,i,in} = h_{f_{FC}}\langle T_{FC,i,in}, X_{FC,i,in}\rangle \tag{4.53.1}$$

$$h_{FC,i,out} = h_{f_{FC}}\langle T_{FC,i,out}, X_{FC,i,out}\rangle \tag{4.53.2}$$

$$h_{V,FC,i} = h_{g}\langle P_{FC,i}, T_{V,FC,i}\rangle = h_{g}\langle P_{FC,i}, T_{FC,i,out}\rangle \tag{4.53.3}$$

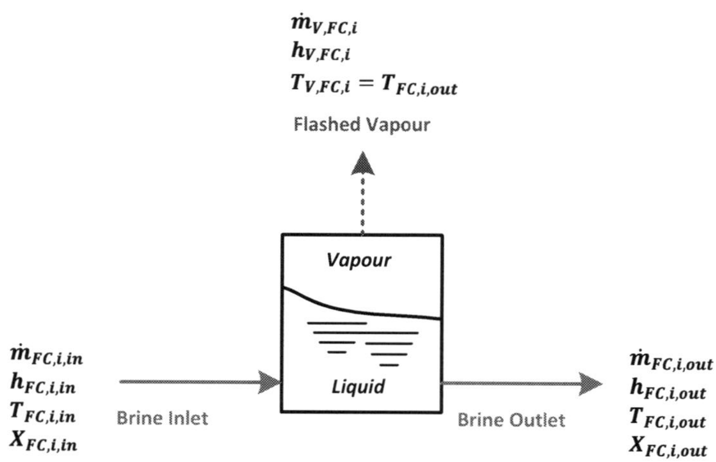

Figure 4.19
Schematic design of a flashing chamber ($i \in \{1,\ldots,j\}$).

Therefore with reference to Eqs. (4.51) and (4.53) the amount of generated flashed vapor is calculated as

$$\dot{m}_{V,FC,i} = \dot{m}_{FC,i,in} \cdot \frac{h_{f_{FC}\langle T_{FC,i,in}, X_{FC,i,in}\rangle} - h_{f_{FC}\langle T_{FC,i,out}, X_{FC,i,out}\rangle}}{h_{g\langle P_{FC,i}, T_{FC,i,out}\rangle} - h_{f_{FC}\langle T_{FC,i,out}, X_{FC,i,out}\rangle}} \tag{4.54}$$

In the previous equations, $P_{FC,i}$ is the pressure of the relevant flashing vessel and

$$T_{FC,i,out} = T_{sat\langle P_{FC,i}\rangle} + BPE_{FC,i} + NEA_{FC,i} \tag{4.55}$$

BPE is the boiling point elevation, and nonequilibrium allowance (*NEA*) for flashing chambers is a function of the range of flashing temperature, saturation temperature, mass flow rate of brine per unit chamber width, brine level inside the flashing chamber, and the design of the flash chamber such as chamber length, width, and orifice type of the brine transfer device from chamber to chamber (see Appendix B) [14]. Under typical operating conditions, *NEA* varies between 0.03°C (as in the first few stages) to 0.8°C (as in the last few low-temperature stages) in a multistage flash desalination plant [15]. In the present simulations for seawater applications, *NEA* can vary from 0.3 (as in the first chamber) to 0.6°C (as in the last chamber), but the results that we present herein pertain to a uniform 0.6°C *NEA* for all the flashing chambers so as to be conservative [5]. For the alumina refinery evaporation plant application (Chapter 9), a fixed 0.5°C *NEA* has been considered [16–18].

For the purpose of simulation and to make a connection between the flashing vessels and the primary MED effects, the pressure of each flashing chamber will be set based on the pressure of the judicious primary MED effect into which the flashed vapor is injected. Therefore a pressure difference (ΔP_{inj}) between 500 Pa and 1 kPa is considered to be sufficient to have a natural flow [3].

$$P_{FC,i} = P_{k^*} + \Delta P_{inj} \tag{4.56}$$

ΔP_{inj} is considered to be 500 Pa and 1.0 kPa for seawater and alumina refinery applications, respectively [4,5,16].

4.2.4.2.3 Brine Recirculation Process

With the aim of decreasing the make-up water flow rate to the flashing section, a brine recirculation configuration is used as shown in Fig. 4.17. For this purpose, a rejected brine stream is then needed to control the mass and salinity balance in the flashing section [19].

The recycling of brine emanating from the series of flashing chambers effectively suppresses the amount of released NCGs and thereby further improves the heat transfer efficacy and attenuates the venting and vacuuming power consumptions [5]. It also can

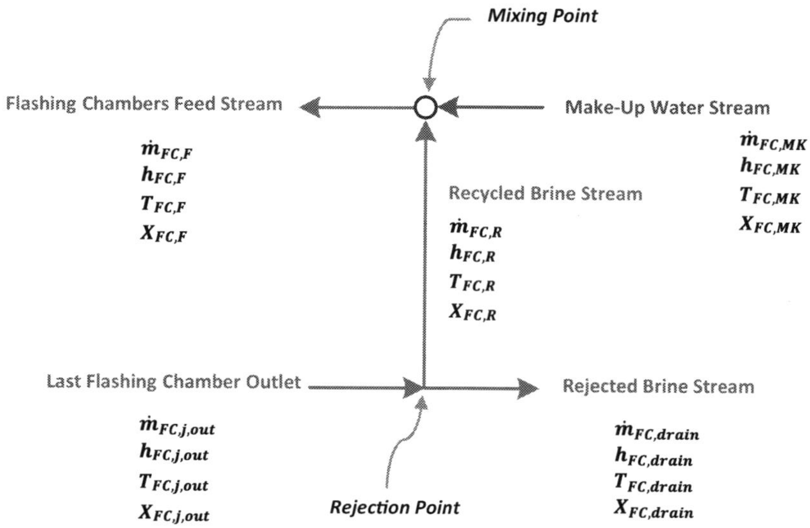

Figure 4.20
Schematic design of the brine recycling process.

decrease the chemical additive consumption and the size of pretreatment and deaeration facilities for the feed stream [19–21].

As shown in Fig. 4.20, in order to compensate for the water loss due to evaporation and brine rejection in the flashing section, a make-up water stream is needed. The mass balance at the mixing point is written as

$$\dot{m}_{\text{FC,MK}} = \dot{m}_{\text{FC},F} - \dot{m}_{\text{FC},R} \tag{4.57}$$

where

$$\dot{m}_{\text{FC},F} = \dot{m}_{\text{c\_hex}} \tag{4.57.1}$$

and from the flashing chambers, we know that

$$\dot{m}_{\text{c\_hex}} = \dot{m}_{\text{FC},j,\text{out}} + \dot{m}_{V,\text{FC,total}} = \dot{m}_{\text{FC},j,\text{out}} + \sum_{i=1}^{j} \dot{m}_{V,\text{FC},i} \tag{4.57.2}$$

For the rejection point, we have

$$\dot{m}_{\text{FC},R} = \dot{m}_{\text{FC},j,\text{out}} - \dot{m}_{\text{FC,drain}} \tag{4.58}$$

where

$$\dot{m}_{\text{FC,drain}} = \dot{m}_{\text{FC},j,\text{out}} \cdot DR \tag{4.58.1}$$

where DR is the drain ratio determined by the solver (see Section 4.2.6). Therefore

$$\dot{m}_{\text{FC},R} = (1 - DR) \cdot \dot{m}_{\text{FC},j,\text{out}} \tag{4.59}$$

which by knowing the *DR* and the outlet mass flow rate from the last flashing chamber, the amount of recycled brine mass flow rate can be calculated.

The salinity balance at the mixing point is written as

$$\dot{m}_{\mathrm{FC,MK}} \cdot X_{\mathrm{FC,MK}} = \dot{m}_{\mathrm{FC},F} \cdot X_{\mathrm{FC},F} - \dot{m}_{\mathrm{FC},R} \cdot X_{\mathrm{FC},R} \tag{4.60}$$

In our seawater application a 70,000 ppm has been considered for $X_{\mathrm{FC},R}$ [4,5,22]. This is the maximum permissible value of the outlet salinity in order to prevent calcium sulfate formation [19]. $X_{\mathrm{FC,MK}}$ is the salinity of the make-up water after deaeration, which to a good approximation can be considered to equate to X_C (see Section 4.2.4.2.4), which in our considered seawater application is 35,000 ppm [4,5,22]. Therefore by knowing the number of flashing chambers and the relevant overall temperature difference across the flashing chambers and with reference to the relevant mass flow rate from Eqs. (4.57) to (4.60) as the relevant boundary conditions, $X_{\mathrm{FC},F}$ can be calculated by the solver.

4.2.4.2.4 Deaerator

As mentioned earlier, the make-up water stream in the brine recirculation system is needed to compensate for water loss due to evaporation and brine rejection. This make-up water should be properly deaerated and vented in order to avoid accumulation of noncondensable gases (NCGs). Otherwise these NCGs are then released in the flashing chambers and transferred to the primary MED effects through the injection lines, which will decrease the heat transfer coefficient in the heat exchanger surfaces and decrease the thermal performance of the process [23]. For the purpose of a proper venting system, a deaerator is needed to remove a portion of these dissolved gasses from the feed seawater inlet to the flashing chamber (Fig. 4.21) so as to attenuate the venting power consumptions. Moreover, deaeration will reduce the oxygen content of the make-up

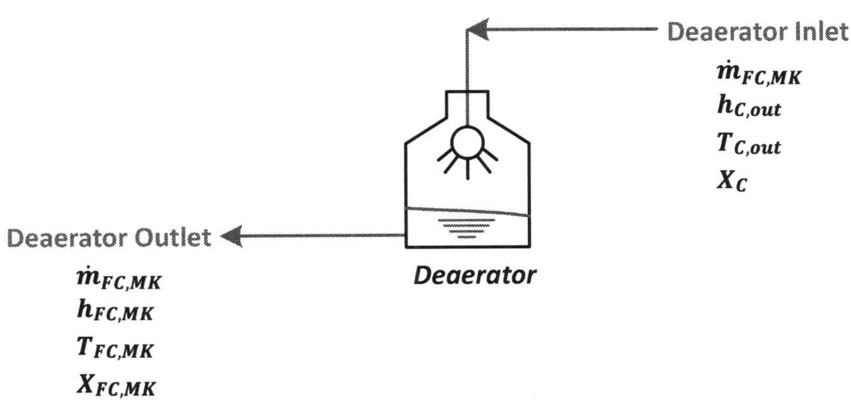

Figure 4.21
Schematic design of the deaerator.

water, which otherwise will cause corrosion problems in the other devices [24]. Nowadays, the deaerator is a rectangular cross-sectioned vessel. It includes an empty space with a lower pressure at the top, where flashing takes place, and a packed section with Pall rings at the bottom, where the stripping process occurs [25]. The feed water is fed from the top to this flashing vessel, and then distributed onto the corrugated plate while the stripping steam is fed from the bottom and comes into contact with the distributed feed in a counter-current direction [24]. An industrial experimental investigation showed the possibility of operating without stripping steam feeding [25], which has been considered in our seawater application.

For this purpose, a 5.9 kPa operating pressure [25] has been considered for the deaeration vessel, which due to the low equivalent temperature drop (2−3°C in our seawater application), to a good approximation the flashing evaporation rate can be ignored in our deaerator [4,5,22]. Therefore there is no mass flow rate change, and the salinity also remains constant (Fig. 4.21). This means that

$$X_{\mathrm{FC,MK}} \approx X_C \tag{4.61}$$

As mentioned previously, the temperature drops and therefore the enthalpy changes from $h_{C,\mathrm{out}}$ to $h_{\mathrm{FC,MK}}$.

$$h_{C,\mathrm{out}} = h_{f_C\langle T_{C,\mathrm{out}}, X_C\rangle} \tag{4.62}$$

$$h_{\mathrm{FC,MK}} = h_{f_{\mathrm{FC}}\langle T_{\mathrm{FC,MK}}, X_C\rangle} \tag{4.63}$$

4.2.5 Overall Mass, Salinity, and Energy Balances

The overall mass, salinity, and energy balances are essential steps in the mathematical simulation. The inlets and outlets of each process have been shown in Figs. 4.22−4.25. The following equations for each process are integral to the simulation.

4.2.5.1 Conventional Multi-Effect Distillation Process

For a conventional MED process (Fig. 4.22) there are two inlets (the main heat source inlet and the cooling water inlet stream) and four outlets (including heat source outlet, freshwater outlet, brine outlet, and cooling water outlet).

The overall mass balance is calculated as

$$\sum \dot{m}_{\mathrm{inlets}} = \sum \dot{m}_{\mathrm{outlets}} \tag{4.64}$$

which for the conventional MED process can be written as

$$\dot{m}_{C,\mathrm{in}} = \dot{m}_{C,\mathrm{out}} + \dot{m}_{B,\mathrm{total}} + \dot{m}_{D,\mathrm{total}} \tag{4.65}$$

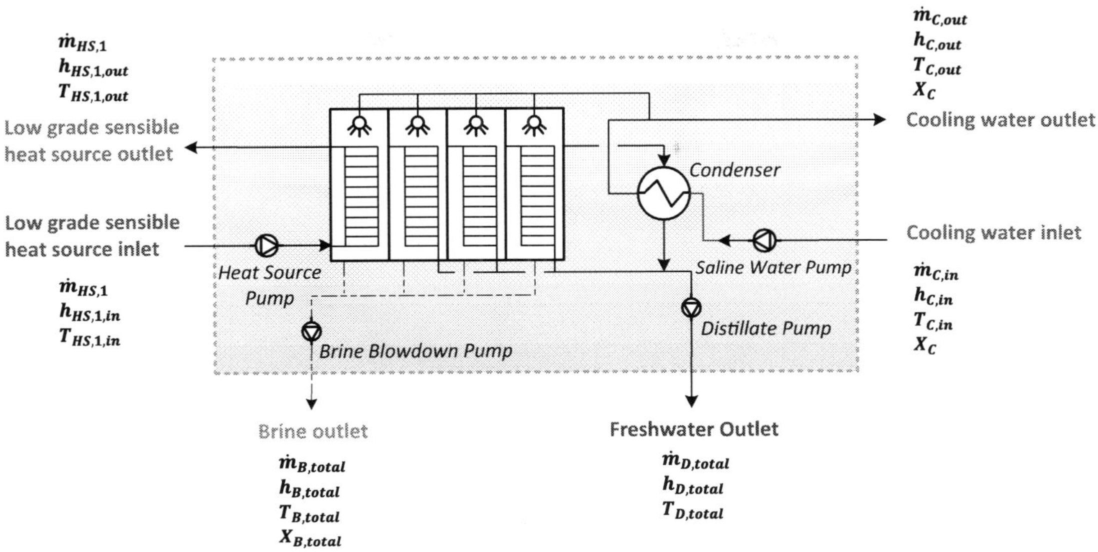

Figure 4.22
Inlets and outlets of a conventional multi-effect distillation process.

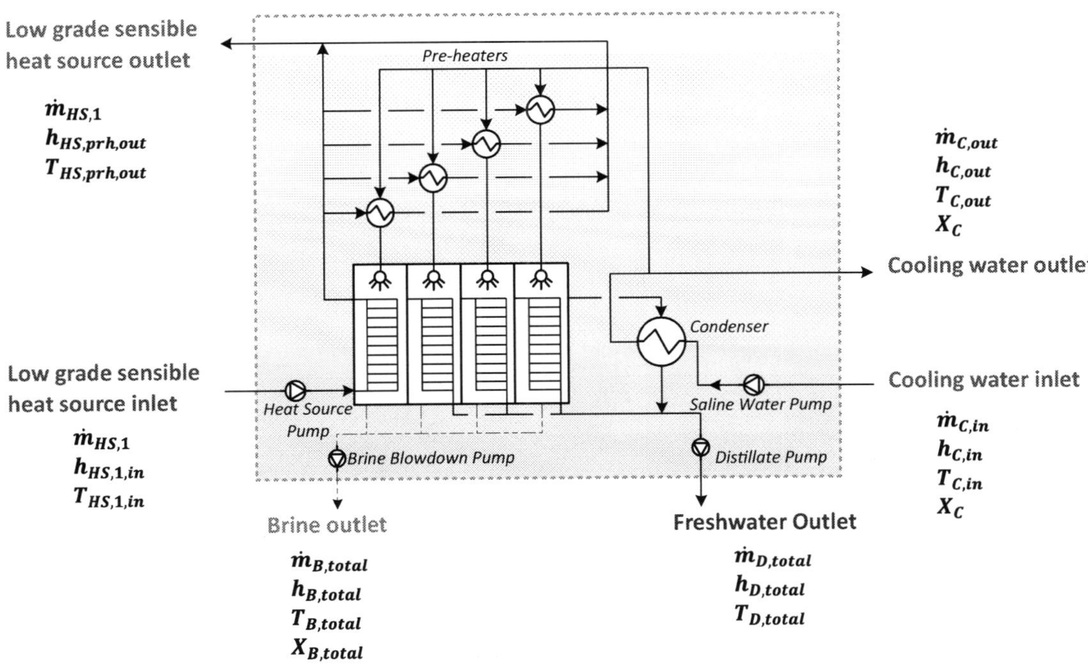

Figure 4.23
Inlets and outlets of a preheated multi-effect distillation process.

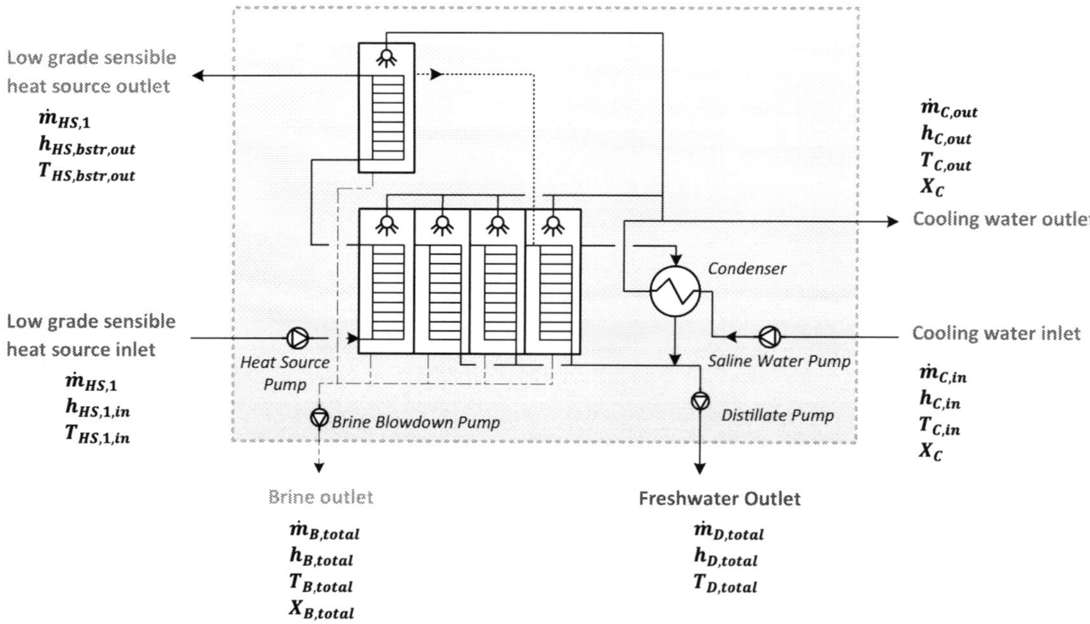

Figure 4.24
Inlets and outlets of a boosted multi-effect distillation process.

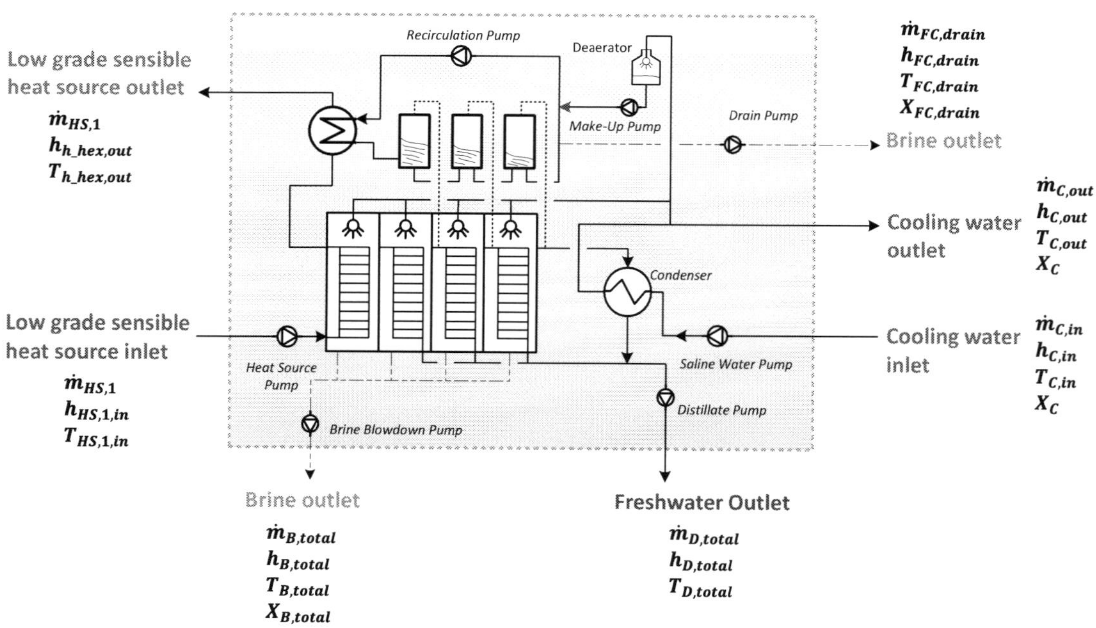

Figure 4.25
Inlets and outlets of a flash-boosted multi-effect distillation process.

where $\dot{m}_{B,\text{total}}$ is the total outlet brine extracted from the primary MED section and $\dot{m}_{D,\text{total}}$ is the total freshwater production rate of the process, which is

$$\dot{m}_{D,\text{total}} = \sum_{k=1}^{n} \dot{m}_{V,k} \tag{4.65.1}$$

and

$$\dot{m}_{C,\text{in}} = \dot{m}_C \tag{4.65.2}$$

$$\dot{m}_{C,\text{out}} = \dot{m}_C - \dot{m}_{F,\text{total}} \tag{4.65.3}$$

The overall salinity balance is

$$\left(\dot{m}_{C,\text{in}} - \dot{m}_{C,\text{out}}\right)\cdot X_C = \dot{m}_{B,\text{total}}\cdot X_{B,\text{total}} \tag{4.66}$$

Finally, for the energy balance we have

$$\sum E_{\text{inlets}} = \sum E_{\text{outlets}} \tag{4.67}$$

$$\begin{aligned}\dot{m}_{\text{HS},1}\cdot\left(h_{\text{HS},1,\text{in}} - h_{\text{HS},1,\text{out}}\right) = \dot{m}_{C,\text{out}}\cdot h_{C,\text{out}} - \dot{m}_{C,\text{in}}\cdot h_{C,\text{in}} + \dot{m}_{B,\text{total}}\cdot h_{B,\text{total}} \\ + \dot{m}_{D,\text{total}}\cdot h_{D,\text{total}}\end{aligned} \tag{4.68}$$

4.2.5.2 Preheated Multi-Effect Distillation Process

For a P-MED process (Fig. 4.23), just as the conventional MED process, there are two inlets (the main heat source inlet and the cooling water inlet stream) and four outlets (including heat source outlet from the preheater section, freshwater outlet, brine outlet, and cooling water outlet). Therefore, Eqs. (4.64)–(4.67) can again be used and Eq. (4.68) is revised to

$$\begin{aligned}\dot{m}_{\text{HS},1}\cdot\left(h_{\text{HS},1,\text{in}} - h_{\text{HS},\text{prh},\text{out}}\right) = \dot{m}_{C,\text{out}}\cdot h_{C,\text{out}} - \dot{m}_{C,\text{in}}\cdot h_{C,\text{in}} + \dot{m}_{B,\text{total}}\cdot h_{B,\text{total}} \\ + \dot{m}_{D,\text{total}}\cdot h_{D,\text{total}}\end{aligned} \tag{4.69}$$

where $h_{\text{HS},\text{prh},\text{out}}$ is the enthalpy of the mixed stream stemming from the preheater outlet hot-side stream and the first effect heat source outlet stream.

4.2.5.3 Boosted Multi-Effect Distillation Process

For a B-MED process (Fig. 4.24), again there are two inlets (the main heat source inlet and the cooling water inlet stream) and four outlets (including heat source outlet from the booster, freshwater outlet, brine outlet, and cooling water outlet). Therefore, Eqs. (4.64)–(4.67) can be used here instead of Eq. (4.65.1), which for a B-MED can be revised to

$$\dot{m}_{D,\text{total}} = \sum_{k=1}^{n} \dot{m}_{V,k} + \dot{m}_{V,\text{bstr}} \tag{4.70}$$

and Eq. (4.68) should also be revised to

$$\dot{m}_{\mathrm{HS},1} \cdot \left(h_{\mathrm{HS},1,\mathrm{in}} - h_{\mathrm{HS,bstr,out}}\right) = \dot{m}_{C,\mathrm{out}} \cdot h_{C,\mathrm{out}} - \dot{m}_{C,\mathrm{in}} \cdot h_{C,\mathrm{in}} + \dot{m}_{B,\mathrm{total}} \cdot h_{B,\mathrm{total}} + \dot{m}_{D,\mathrm{total}} \cdot h_{D,\mathrm{total}} \tag{4.71}$$

where $h_{\mathrm{HS,bstr,out}}$ is the heat source outlet stream from the booster.

4.2.5.4 Flash-Boosted Multi-Effect Distillation Process

For an FB-MED process (Fig. 4.25) there are two inlets (the main heat source inlet and the cooling water inlet stream) and five outlets (including heat source outlet, freshwater outlet, brine outlets from the primary MED and flashing chambers, and cooling water outlet).

The overall mass balance based on Eq. (4.64) is written as

$$\dot{m}_{C,\mathrm{in}} = \dot{m}_{C,\mathrm{out}} + \dot{m}_{B,\mathrm{total}} + \dot{m}_{\mathrm{FC,drain}} + \dot{m}_{D,\mathrm{total}} \tag{4.72}$$

where $\dot{m}_{B,\mathrm{total}}$ is the total outgoing brine extracted from the primary MED section; $\dot{m}_{\mathrm{FC,drain}}$ is the rejected brine from the flashing section; $\dot{m}_{D,\mathrm{total}}$ is the total freshwater production rate of the process, which is

$$\dot{m}_{D,\mathrm{total}} = \sum_{k=1}^{n} \dot{m}_{V,k} + \sum_{i=1}^{j} \dot{m}_{V',i} \tag{4.72.1}$$

and

$$\dot{m}_{C,\mathrm{in}} = \dot{m}_{C} \tag{4.72.2}$$

$$\dot{m}_{C,\mathrm{out}} = \dot{m}_{C} - \dot{m}_{F,\mathrm{total}} - \dot{m}_{\mathrm{FC,MK}} \tag{4.72.3}$$

The overall salinity balance is

$$\left(\dot{m}_{C,\mathrm{in}} - \dot{m}_{C,\mathrm{out}}\right) \cdot X_C = \dot{m}_{B,\mathrm{total}} \cdot X_{B,\mathrm{total}} + \dot{m}_{\mathrm{FC,drain}} \cdot X_{\mathrm{FC,drain}} \tag{4.73}$$

Finally for the energy balance we have

$$\sum E_{\mathrm{inlets}} = \sum E_{\mathrm{outlets}} \tag{4.74}$$

$$\dot{m}_{\mathrm{HS},1} \cdot \left(h_{\mathrm{HS},1,\mathrm{in}} - h_{h_{\mathrm{hex}},\mathrm{out}}\right) = \dot{m}_{C,\mathrm{out}} \cdot h_{C,\mathrm{out}} - \dot{m}_{C,\mathrm{in}} \cdot h_{C,\mathrm{in}} + \dot{m}_{\mathrm{FC,drain}} \cdot h_{\mathrm{FC,drain}} + \dot{m}_{B,\mathrm{total}} \cdot h_{B,\mathrm{total}} + \dot{m}_{D,\mathrm{total}} \cdot h_{D,\mathrm{total}} \tag{4.75}$$

4.2.6 Solving Procedure

The general solving procedure flowchart is shown in Fig. 4.26. A GRG method [2] is used to solve all the previous equations by imposing the boundary conditions from the laws of

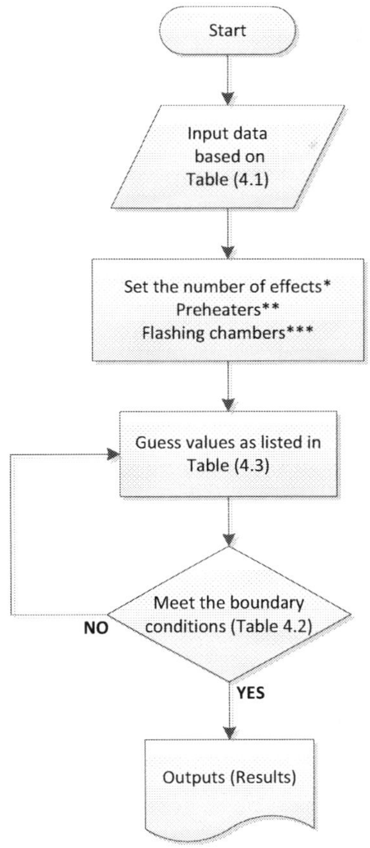

Figure 4.26
The relevant flowchart of the process simulation (*Applicable to all configurations, **Applicable to P-MED configuration, ***Applicable to FB-MED configuration).

thermodynamics and appropriate operational, technical, and economic constraints as explained in the following chapters [4,5,16,22].

The assumptions of our process simulation are listed in Table 4.2. By using these assumed parameters, and those relevant boundary conditions listed in Table 4.3, which will be described further in Chapters 8 and 9, the solver (Appendix G) by guessing some essential parameters (Table 4.4) will produce the results such as the total freshwater production rate $\left(\dot{m}_{D,\text{total}}\right)$; total concentrated feed flow rate (as brine in seawater application), which is important in mineral refining processes; waste heat performance ratio (PR_{WH}); temperature profiles across the effects; and UA values. These technical results are then used for the thermoeconomic evaluation as explained in Chapter 7.

Table 4.2: Assumed thermodynamic parameters for the solving procedure (input data).

Parameter	Description
$T_{HS,1,in}$	Heat source inlet temperature (°C)
$\dot{m}_{HS}$	Heat source mass flow rate (kg/s)
$T_{C,in}$	Cooling water inlet temperature (°C)
$X_{C,in}$	Feed water salinity (ppm)
R	Feed-to-vapor ratio
TBT	Top brine temperature °C), which is 70°C for multi-effect distillation process
ΔP_{inj} [a]	Pressure difference for steam injection purpose (Pa)
$\min \Delta T_e$	Minimum temperature difference across each multi-effect distillation effect (°C)
$\min \Delta T_{HS,1}$	Minimum heat source temperature drop across the first effect (°C)
$\min \Delta T_{hex}$ [b]	Minimum temperature drop across liquid–liquid heat exchanger (°C)
$\Delta T_{approach}$ [b]	Liquid–liquid heat exchanger approach temperature (°C)
$\min \Delta T_C$	Minimum temperature difference across the condenser (°C)
X_{max}	Maximum allowable outlet salinity (ppm), which is 70,000 (ppm) in seawater application
$X_{FC,j,out}$	Flashing section brine outlet salinity (ppm)
$\dot{m}_{vent}$	Venting flow rate, which is considered as a percentage of $\dot{m}_{HS,cond}$ (see Chapter 5)
$P_{deaerator}$ [c]	The deaerator operating pressure

[a] Applicable for boosted and flash-boosted multi-effect distillation processes.
[b] Applicable for preheated and flash-boosted multi-effect distillation processes.
[c] Applicable for flash-boosted multi-effect distillation process.

Table 4.3: Boundary conditions.

Configuration	Boundary Condition
All configurations	$\min\Delta T_{HS,1} \leq (T_{HS,1,in} - T_{HS,1,out})$
	$T_{B,1} \leq T_{pinch,1}$
	$T_{B,1} \leq \text{TBT}$
	$T_{C,out} \leq \left(T_{sat\langle P_n \rangle} - \min\Delta T_e\right)$
	$X_{B,k} \leq X_{max}$
P-MED	$\Delta T_{approach} \leq (T_{h\_prh,i,in} - T_{c\_prh,i,out})$
	$\Delta T_{approach} \leq (T_{h\_prh,i,out} - T_{c\_prh,i,in})$
	$\min \Delta T_{hex} \leq (T_{h\_hex,i,in} - T_{h\_hex,i,out})$
	$\min \Delta T_{hex} \leq (T_{c\_hex,i,out} - T_{c\_hex,i,in})$
B-MED	$\min \Delta T_{HS,1} \leq (T_{HS,bstr,in} - T_{HS,bstr,out})$
	$T_{B,bstr} \leq T_{pinch,bstr}$
	$T_{B,bstr} \leq \text{TBT}$
	$T_{sat\langle P_{k-1}+P_{inj} \rangle} \leq T_{sat\langle P_{bstr} \rangle}$; k is the injected effect number
	$X_{B,bstr} \leq X_{max}$
FB-MED	$\min \Delta T_{hex} \leq (T_{h\_hex,in} - T_{h\_hex,out})$
	$\min \Delta T_{hex} \leq (T_{c\_hex,out} - T_{c\_hex,in})$
	$X_{FC,j,out} \leq X_{max}$

Table 4.4: Guessed parameters.

Value	Relevant Configuration
$T_{HS,1,out}$	All configurations (MED, P-MED, B-MED, FB-MED)
$T_{B,1}$	All configurations (MED, P-MED, B-MED, FB-MED)
$T_{h\_prh,i,out}$	P-MED
$T_{HS,bstr,out}$	B-MED
$T_{B,bstr}$	B-MED
$T_{h\_hex,out}$	FB-MED
$X_{FC,1,in}$	FB-MED
$X_{FC,i,out}$	FB-MED
DR	FB-MED

B-MED, boosted multi-effect distillation; *FB-MED*, flash-boosted multi-effect distillation; *P-MED*, preheated multi-effect distillation; *MED*, multi-effect distillation.

References

[1] ALFALAVAL, Single Effect Freshwater Generator, Model JWP-16/26-C Series, in Alfa Laval Marine & Diesel Product Catalogue, Alfa Laval Corporate AB, 2003.

[2] L.S. Lasdon, A.D. Waren, A. Jain, M. Ratner, J. Rice, Design and testing of a generalized reduced gradient code for nonlinear programming, ACM Trans. Math. Softw. 4 (1) (1978) 34–50.

[3] A. Christ, K. Regenauer-Lieb, H.T. Chua, Boosted multi-effect distillation for sensible low-grade heat sources: a comparison with feed pre-heating multi-effect distillation, Desalination 366 (2015) 32–46.

[4] B. Rahimi, J. May, A. Christ, K. Regenauer-Lieb, H.T. Chua, Thermo-economic analysis of two novel low grade sensible heat driven desalination processes, Desalination 365 (2015) 316–328.

[5] B. Rahimi, A. Christ, K. Regenauer-Lieb, H.T. Chua, A novel process for low grade heat driven desalination, Desalination 351 (October 2014) 202–212.

[6] NIST, NIST Reference Fluid Thermodynamic and Transport Properties Database (REFPROP): Version 9.1, 2013.

[7] M.H. Sharqawy, J.H. Lienhard V, S.M. Zubair, Thermophysical properties of seawater: a review of existing correlations and data, Desalin. Water Treat. 16 (10) (2010) 354–380.

[8] J.W. Bertetti, W.L. McCabe, Specific heats of sodium hydroxide solutions, Ind. Eng. Chem. 28 (3) (1936) 375–378.

[9] J.W. Bertetti, W.L. McCabe, Sodium hydroxide solutions, Ind. Eng. Chem. 28 (2) (1936) 247–248.

[10] H.R. Wilson, W.L. McCabe, Specific heats and heats of dilution of concentrated sodium hydroxide solutions, Ind. Eng. Chem. 34 (5) (1942) 558–566.

[11] W.L. McCabe, The enthalpy – concentration chart – a useful device for chemical engineering calculations, Trans. A.I.Ch.E 31 (1935) 129–169.

[12] E.W. Washburn, C.J. West, C. Hull, National Academy of Sciences (U.S.), International Council of Scientific Unions, and National Research Council (U.S.), International Critical Tables of Numerical Data, Physics, Chemistry and Technology, vol. III, McGraw-Hill, New York, 1928.

[13] Solvay Chemicals Co., Liquid Caustic Soda Characteristics, 2014 [Online]. Available: http://www.solvaychemicals.com/EN/products/causticsoda/Liquidcausticsoda.aspx.

[14] H. El-Dessouky, H.I. Shaban, H. Al-Ramadan, Steady-state analysis of multi-stage flash desalination process, Desalination 103 (1995) 271–287.

[15] P. Fiorini, E. Sciubba, C. Sommariva, A new formulation for the non-equilibrium allowance in MSF processes, Desalination 136 (2001) 177–188.

[16] B. Rahimi, K. Regenauer-Lieb, H.T. Chua, E. Boom, S. Nicoli, S. Rosenberg, A novel low grade heat driven process to re-concentrate process liquor in alumina refineries, in: 10th Int. Alumina Quality Workshop (AQW) Conference, Perth, Australia, 19th–23rd April, 2015, pp. 327–336.

[17] B. Rahimi, K. Regenauer-Lieb, H.T. Chua, E. Boom, S. Nicoli, S. Rosenberg, A novel low grade heat driven process to re-concentrate process liquor in alumina refineries, Hydrometallurgy (2015) 327–336.

[18] B. Rahimi, K. Regenauer-Lieb, H.T. Chua, E. Boom, S. Nicoli, S. Rosenberg, A novel flash boosted evaporation process for alumina refineries, Appl. Therm. Eng. 94 (2016) 375–384.

[19] H.T. El-Dessouky, H.M. Ettouney, Fundamentals of Salt Water Desalination, Elsevier Science B.V., 2002.

[20] A.M. Helal, M. Odeh, The once-through MSF design. Feasibility for future large capacity desalination plants, Desalination 166 (August 2004) 25–39.

[21] A.M. Helal, Once-through and brine recirculation MSF designs a comparative study, Desalination 171 (2004) 33–60.

[22] B. Rahimi, K. Regenauer-Lieb, H.T. Chua, A novel desalination design to better utilise low grade sensible waste heat resources, in: IDA World Congress 2015 on Desalination and Water Reuse, San Diego, US, 30-August to 4-September 2015.

[23] R. Semiat, Y. Galperin, Effect of non-condensable gases on heat transfer in the tower MED seawater desalination plant, Desalination 140 (1) (2001) 27–46.

[24] P. Costa, A. Ferro, E. Ghiazza, B. Bosio, Seawater deaeration at very low steam flow rates in the stripping section, Desalination 201 (1–3) (2006) 306–314.

[25] E. Ferro, E. Ghiazza, B. Bosio, P. Costa, Modelling of flash and stripping phenomena in deaerators for seawater desalination, Desalination 142 (2) (2002) 171–180.

CHAPTER 5

Pumping Power Analysis

5.1 Introduction

The electrical power consumption is one of the most important considerations in the design of MED systems. It stems primarily from the pumping power consumption in desalination plants. The pumping power is calculated according to Eq. (5.1) [1].

$$Pumping\ Power(\mathrm{kW}) = \frac{\Delta P(\mathrm{kPa})\cdot \dot{V}(\mathrm{m^3/s})}{\eta_{pump}\cdot \eta_{motor}} \tag{5.1}$$

where η_{pump} and η_{motor} are the pump and motor efficiencies and equate to 0.7 and 0.9, respectively, in our present simulation; however, these values are not fixed for a given pump.

In desalination plants a specific pumping power is used instead of net pumping power. This parameter shows the amount of electrical energy that a desalination plant is used to produce one unit of freshwater (Eq. (5.2)). Namely,

$$p(\mathrm{kWh/m^3}) = \frac{Pumping\ Power(\mathrm{kW})}{\dot{V}_{D,total}(\mathrm{m^3/h})} \tag{5.2}$$

where p is the specific pumping power in (kWh/m^3) and $\dot{V}_{D,total}$ is the total production rate of the plant in (m^3/h).

As schematically shown in Figs. 2.2–2.5, in conventional, preheated (P-), boosted (B-), and flash-boosted (FB-) multi-effect distillations (MEDs), the pumping power for those relevant pumps for each of those inlet and outlet streams will be discussed in the following sections [1].

5.2 Pressure Drops in Desalination Plants

As shown in Eq. (5.1), the pressure drop is an important parameter of pumping power consumption. This parameter depends on the process design constraints and the pumping cost. As we know, the pumps are used to drive the fluids (namely, the heat source medium, freshwater, feed water, brine, and noncondensable gasses in desalination processes) through the process equipment such as pipes, vessels, heat exchangers, and so on. For this purpose, they should compensate for the pressure drop that occurs in each type of equipment and also provide adequate pressure for conveyance.

Low Grade Heat Driven Multi-Effect Distillation and Desalination. http://dx.doi.org/10.1016/B978-0-12-805124-5.00005-X

In low grade heat-driven desalination plants, the major pumps are those that transport the fluids through the heat exchangers; namely, saline water pump, brine recirculation pump (which is only applicable in FB-MED processes), and heat source medium pump (see Figs. 2.2–2.5). The required energy to transport the relevant medium through heat exchangers is a function of the heat transfer coefficient. Thus, in the case of using a properly sized pump to provide for an adequate pressure, a higher heat transfer coefficient leads to a smaller heat exchanger. In other words, in order to decrease the heat exchanger area, which translates to a lower capital cost, we need to use a powerful pump, which then increases the operating cost. Therefore both the capital and operating costs should be considered so as to reach an optimum economic design for selecting the proper heat exchanger and its relevant pump.

Following from this explanation, the pressure drop depends on the type of heat exchanger. In this book a plate-type heat exchanger and evaporator have been considered and the calculation of the pressure drop in these types of heat exchangers will be explained in Appendix C. We need to mention that horizontal falling film evaporators are also very common in MED plants in the desalination industry. Due to the different structure of these types of evaporator as compared to the plate types, the pressure drop characteristics will be different.

The major pumps that are used in low grade sensible heat-driven desalination plants as shown in Figs. 2.2–2.5 are described next.

5.2.1 Saline Water Pump

This pump conveys saline water through the condenser of the primary MED and feeds the MED effects. In the P-MED scheme, it should provide for the cold side pressure drop of the preheaters. In our simulations, a plate-type condenser has been considered and its pressure drop can be calculated based on the relevant equations in Appendix C. The overall pressure head for this pump is taken to be the pressure drop across the condenser (Appendix C) plus 1.0 bar to access the atmospheric pressure [1].

$$\Delta P_{SWP} = 1.0 + \Delta P_{cond} \tag{5.3}$$

5.2.2 Brine Recirculation Pump

This pump is used in the FB-MED system to recirculate the brine through the brine heater (namely, a plate-type liquid–liquid heat exchanger) and the flashing chambers. As shown in Fig. 2.5, the major duty of this pump is the plate-type brine heater pressure drop, and the pressure head is calculated according to Eq. (5.4). For this purpose, the plate-type brine heater's pressure drop has been calculated according to the procedure in Appendix C and the germane pressure head is calculated based on the flashing chamber sections' inlet and outlet saturated pressure difference. Therefore:

$$\Delta P_{BRP} = \Delta P_{hex} + \left(P_{FC,1,in} - P_{FC,j,out}\right) \tag{5.4}$$

5.2.3 Heat Source Medium Pump

In a conventional MED process, this pump is used for pumping the heat source medium through the first effect. In a P-MED process (Fig. 2.3), in addition to the first effect, the pump needs to compensate for the pressure drop at the hot side of the preheaters. For B-MED as shown in Fig. 2.4, the pressure drop of the booster (which is effectively equivalent to another effect) should be considered. In an FB-MED (Fig. 2.5), this pump is used for pumping through the first effect of the primary MED plant and the liquid–liquid heat exchanger that is used as the brine heater in the flashing section.

In all these configurations, the pressure drop in the first effect, relevant heat exchangers, and the booster should be calculated based on the types of evaporator and heat exchanger. In this book a plate-type evaporator and heat exchanger have been considered. The pressure drop of these kinds of heat exchangers are explained in Appendix C.

For a conventional MED process,

$$\Delta P_{HSP} = \Delta P_{HS,1} \tag{5.5}$$

where $\Delta P_{HS,1}$ is the pressure drop through the hot side of the first-effect evaporator.

For a P-MED process,

$$\Delta P_{HSP} = \Delta P_{HS,1} + \Delta P_{HS,prh} \tag{5.6}$$

where $\Delta P_{HS,prh}$ is the pressure drop of the hot side of preheaters. In the case of using a portion of the main heat source stream in the preheater section (as shown in Fig. 2.3), it may be better to use a secondary pump to compensate for the preheaters' pressure drop specifically for that line.

For a B-MED process,

$$\Delta P_{HSP} = \Delta P_{HS,1} + \Delta P_{HS,bstr} \tag{5.7}$$

For an FB-MED process,

$$\Delta P_{HSP} = \Delta P_{HS,1} + \Delta P_{HS,hex} \tag{5.8}$$

where $\Delta P_{HS,hex}$ is the pressure drop of the brine heater (which in our case is a plate-type liquid–liquid heat exchanger) on its heat source (hot) side.

5.2.4 Distillate Extraction and Brine Blow-Down Pumps

These pumps are used to extract the distillate (freshwater) and brine from the primary MED plant in all the configurations. For these pumps a 2.0 bar differential pressure is considered sufficient, which translates to 1.0 bar reach from vacuum to the atmospheric pressure and 1.0 bar to transfer to an access point [1].

$$\Delta P_{DEP} = \Delta P_{BBP} = 2.0 \tag{5.9}$$

5.2.5 Drain Pump

This pump is only applicable for the FB-MED process. A drain pump is used for brine rejection from the brine recirculation stream of the flashing section. The pumping power of this stream is negligible because of the small flow rate. However, a 2.0 bar differential pressure is considered for this pump for the pumping from vacuum to atmospheric pressure and to reach an access point [1].

$$\Delta P_{DRP} = 2.0 \tag{5.10}$$

5.2.6 Make-Up Water Pump

This pump is only applicable in the flashing section of the FB-MED configuration. As shown in Fig. 2.5, it is used to provide sufficient pressure to the make-up water stream that exits the deaerator. A 2.0 bar differential pressure has been considered for this pump, which includes 1.0 bar to access the atmosphere from vacuum and 1.0 bar for mixing with the recirculated brine that is flowing through the brine recirculation stream [1].

$$\Delta P_{MKP} = 2.0 \tag{5.11}$$

5.2.7 Noncondensable Gas Extraction Vacuum Pump

One of the important issues for low grade sensible heat-driven desalination processes is related to the venting process, which serves to remove the generated noncondensable gases (NCGs) and the air leakage into the system. In the steam-driven desalination process, such as thermal vapor compression MED or multi-stage flash plants, steam ejectors are used for vacuuming and NCG removal, but in low grade sensible heat-driven applications there is no pressurized steam for driving ejectors, and so two options are available. The first option is to use water eductors or air ejectors, and the second is to use water-ring vacuum pumps. In the present work, the second option is adopted [1–3] and the corresponding pumping power consumption is evaluated according to the manufacturer's catalogs for water-ring vacuum pumps [4]. The pumping power of this type of vacuum pump is a function of NCG mass flow rate and the pressure of the vessel under vacuum.

The amount of NCG released is a function of feed water composition, evaporator design, and working conditions. Literature expounding the methods for estimating the amount of NCGs abound [5–7], but in terms of design, it is typical to consider an overdesigned venting system for handling the air leakage into the system [8]. As a matter of comparing among all configurations (MED, P-MED, B-MED, and FB-MED), a cascaded venting scheme is considered for all systems, which means that the NCGs are extracted from each effect and cascaded downstream and eventually into the condenser where all accumulated NCGs are then extracted by the vacuum pump.

In FB-MED processes, because of the low top brine temperature and the make-up/brine recirculation ratio of the present flashing chambers and also the use of deaerated make-up water, the amount of generated NCGs in the present flashing chambers as compared to the released NCGs in the primary MED effects is relatively low. In the present simulation the amount of NCGs released in the flashing chambers is considered based on the ratio between the make-up water of the flashing chambers and the feed water of the primary MED plant that is varied between 15% and 25% (for seawater application) [1–3]. Therefore one venting system is considered for handling all NCGs released from both the primary MED section and flashing chambers. Following routine considerations the venting system is designed to remove 1% of the vapor in the condenser as NCGs [9] in all the configurations.

References

[1] B. Rahimi, A. Christ, K. Regenauer-Lieb, H.T. Chua, A novel process for low grade heat driven desalination, Desalination 351 (October 2014) 202–212.

[2] B. Rahimi, J. May, A. Christ, K. Regenauer-Lieb, H.T. Chua, Thermo-economic analysis of two novel low grade sensible heat driven desalination processes, Desalination 365 (2015) 316–328.

[3] B. Rahimi, K. Regenauer-Lieb, H.T. Chua, A novel desalination design to better utilise low grade sensible waste heat resources, in: IDA World Congress 2015 on Desalination and Water Reuse, San Diego, US, August 30–September 4, 2015.

[4] S.R.L. Gardner Denver, Robuschi Liquid Vacuum Pumps Catalogue, 2016 (Online). Available: http://www.gardnerdenver.com/robuschi/downloads/vacuum-pumps/.

[5] K. Genthner, A. Seifert, A calculation method for condenser in multi-stage evaporators with non-condensable gases, Desalination 81 (1991) 349–366.

[6] H. Glade, J. Meyer, S. Will, The release of CO_2 in MSF and ME distillers and its use for the recarbonation of the distillate: a comparison, Desalination 182 (2005) 99–110.

[7] A.E. Al-Rawajfeh, H. Glade, H.M. Qiblawey, J. Ulrich, Simulation of CO_2 release in multiple-effect distillers, Desalination 166 (2004) 41–52.

[8] M.A. Darwish, M.M. El-Refaee, M. Abdel-Jawad, Developments in the multi-stage flash desalting system, Desalination 100 (1995) 35–64.

[9] A. Seifert, K. Genthner, A model for stagewise calculation of non-condensable gases in multi-stage evaporators, Desalination 81 (1991) 333–347.

CHAPTER 6

Waste Heat Performance Ratio

6.1 Introduction

For most of the current desalination applications, fuel costs are proportional to the energy input to the system. These costs can either be of a direct nature, for example where fossil fuels are used to drive the system, or of an indirect nature, when for example low pressure steam is extracted from a power plant turbine. Consequently, the rational definition of the efficiency of these systems is related to the specific energy consumption per produced freshwater. This is exemplified by the industrial standard gain output ratio (*GOR*) or the performance ratio (*PR*) for steam-driven multi-effect distillation (MED) systems.

In steam-driven processes, the latent heat of steam is utilized as the driving force, the amount of which is calculated by knowing the steam flow rate at the prevailing steam temperature and pressure. In an optimized steam-driven process with a known amount of available latent heat energy, the maximum number of effects (which is limited by the top brine temperature, cooling water temperature, and design temperature differential across each effect) brings about the maximum production rate. Therefore, in each case (with the same steam operational condition that includes temperature, pressure, and flow rate) any improvement can be benchmarked against the amount of steam used under the same process conditions (i.e., same steam operating condition, number of effects, and production rate). In this situation, less steam consumption per unit of freshwater means a more thermally efficient plant.

But in low-grade sensible heat-driven desalination processes where the heat source outlet temperature is changing by increasing the number of effects, an optimized number of effect and heat source temperature drop will be achieved. For this case, the conventional performance parameters such as *GOR* and *PR* are not suitable to benchmark the performance of these systems and a modified parameter is necessary to address the performance of such processes, which are coupled with sensible heat sources.

6.2 Performance of Conventional Steam-Driven Multi-Effect Distillation Process

In conventional steam-driven systems such as multi-stage flash or thermal vapor compression MED, the *GOR*, Eq. (6.1), which is the ratio of the freshwater production rate

Low Grade Heat Driven Multi-Effect Distillation and Desalination. http://dx.doi.org/10.1016/B978-0-12-805124-5.00006-1

to the heat source steam input or the PR, Eq. (6.2), facilitates a quick and effective benchmark for comparing these systems.

$$GOR = \frac{\dot{m}_D}{\dot{m}_{HS,steam}} \tag{6.1}$$

where $\dot{m}_{HS,steam}$ is the mass flow rate of steam supplied to the system, and $\dot{m}_D$ is the mass flow rate of the distillate. The *GOR* is a simplification for steam-driven systems, where the difference between the latent heat of condensation of the heat source medium and the evaporation energy of the feed can be neglected. According to Eq. (6.1), any improvement on steam-driven systems can be interpreted as the use of less steam to produce one unit of production rate.

The *PR* is a far more general benchmark, where the amount of energy used for the evaporation of freshwater is compared against the initial energy input.

$$PR = \frac{\dot{m}_D \cdot \Delta h_{ref}}{\dot{m}_{HS} \cdot \left(h_{HS,in} - h_{HS,out}\right)} \tag{6.2}$$

where Δh_{ref} is equal to 2336 kJ/kg as an industrial benchmark for the specific reference enthalpy of the distillate [1,2].

Since the bulk of the initial input energy is transferred as latent heat from effect to effect, only a relatively small amount of sensible heat is rejected with the brine and freshwater streams from each effect. Therefore whereas the energy is downgraded in terms of temperature with each effect, each of the subsequent processes issues only a small attrition. Consequently the freshwater production in each downstream effect is only slightly reduced, and the production rate can nearly increase proportionately with the number of effects.

Ergo, in order to achieve a high production rate, it is typical for steam-driven systems to install as many effects as possible. The maximal possible number of effects is typically limited by the following three factors:

1. The top brine temperature (TBT) that is the highest temperature the seawater can reach in the system, which is typically the temperature of the brine in the first effect. It is imposed by either the heat source temperature or material limitations to avoid excessive scaling and corrosion. For seawater applications, this maximum temperature is commonly set to between 62°C and 75°C [2,3], depending on the local seawater composition.
2. The cooling water temperature, which is the lowest temperature in the system.
3. The design temperature differential over each effect.

Since the first two factors are constrained by operational conditions, minimizing the design temperature differential is the only means that eventuates a *PR* increase. Because a reduction in temperature differential is associated with increased heat exchanger size and costs, this factor is typically limited by economic considerations.

A commonly adopted temperature differential for conventional systems is around 3°C. Nonetheless, with a cost-optimized material selection and heat exchanger design, temperature drops as low as 1.5–2.5°C can be achieved for commercial applications [4].

Referring to typical operating conditions, *PR*s between 6 and about 12 are emblematic for steam-driven MED systems. Higher *PR*s up to around 17 are possible, especially with additional system augmentation like thermal vapor compression, but this is achieved with a significantly increased capital cost [5].

6.3 Performance of Sensible Heat-Driven Multi-Effect Distillation Process

A disparate design paradigm arises when there is no strong relationship between the energy costs and the energy consumed by the system. This includes a broad range of waste heat sources where hot fluid streams are rejected unused to the environment, often even the purging of which requires freshwater and auxiliary energy. Also, many types of renewable energy subsume to this category. One prime example is geothermal energy, where the costs incurred for the provision of the heat source, namely the construction of the geothermal well, and the pumping of the hot fluid to the surface, are independent of how much energy is finally extracted from that groundwater.

It is evident that for this type of energy, benchmarking a system solely by means of the minimal energy consumption per unit of freshwater is irrational. Instead it is far more worthwhile to exploit the heat sources to bring about the maximal possible freshwater production. This distinction is particularly important for sensible heat sources for the following reason.

In a steam-driven MED plant the latent heat released by the phase change of the heating medium is utilized as the driving force (ignore the desuperheating process). Therefore, the major share of enthalpy of the heating medium is already exploited by the condensation. Since this is an isothermal process, the top brine temperature can closely approach the heat source input temperature, regardless of how much latent heat is extracted (Fig. 6.1).

In contrast, sensible heat sources ordain a temperature drop in order to provide energy; notwithstanding, the separation of the feed seawater still requires a partial phase change. In these cases where low-grade sensible heat source is used instead of steam, a conventional MED process is accompanied by a temperature drop for the heat source stream. Accordingly, the maximal possible top brine temperature is dictated by the gradient of that temperature drop, rather than the heat source input temperature per se as

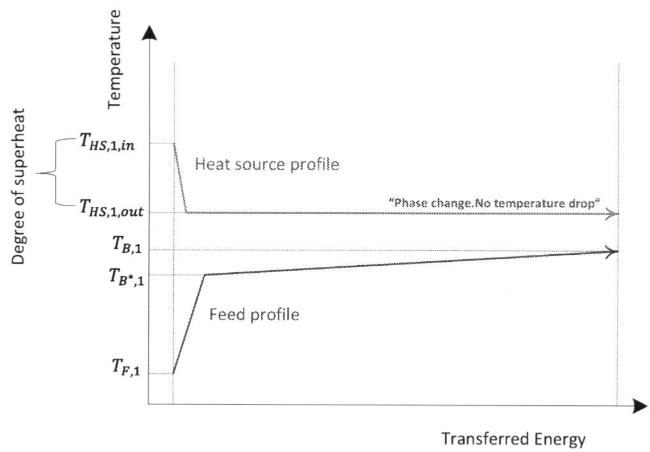

Figure 6.1
Temperature-energy profile for the first effect of a steam-driven multi-effect distillation.

resulting from the pinch point analysis (Fig. 6.2). (See Chapter 4 for more details regarding the feed temperature profile).

In these processes, the outlet heat source temperature and the amount of released energy vary according to the heat source inlet temperature, which shapes the optimization of the conventional MED process. In this context, having a higher number of effects does not guarantee higher production rates [6]. This is because with the same process conditions (such as having the same cooling water temperature, top brine temperature, and design parameters), a higher number of effects means higher heat source outlet temperature,

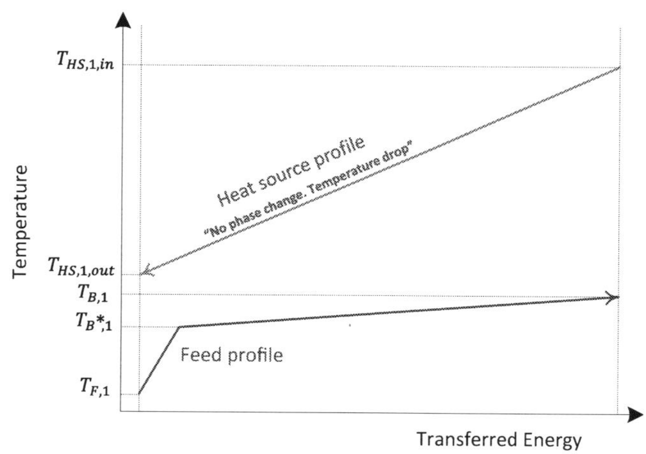

Figure 6.2
Temperature—energy profile for the first effect of a sensible heat-driven multi-effect distillation (see Chapter 4 for more details).

which means a lower temperature drop in the heat source stream and a smaller amount of energy extraction at the first effect. In contrast, a lesser number of effects produce a higher temperature drop and larger energy extraction from the heat source, which suppress the maximal possible TBT, but the production rate per effect cannot then be multiplied through a greater number of effects because this in turn weighs against the achievable number of effects, and the performance of the system can suffer correspondingly. Therefore, for the purpose of optimization, a balance between the number of effects and heat source temperature drop is always necessary for sensible heat-driven processes (with respect to the operational and design limitations that include top brine and cooling water temperatures and temperature differential across each effect [1]) so as to find the configuration that maximizes the production rate.

6.4 The Waste Heat Performance Ratio

Referring to the aforementioned issues, insofar as low-grade heat applications are concerned, the *PR* that holds that heat comes with a premium as it is consumed, does not capture the essence of a desalination system driven by such heat sources, which only attracts a one-off investment cost, instead of an ongoing cost of thermal energy used. In other words, for this type of heat source, benchmarking the system simply by means of the minimal heat source energy consumption per unit production rate is irrational [1]. It is far better to exploit the heat source as much as possible (insofar as operational and design limitations permit) to bring about maximal production rate [1]. To this end, a modified *PR* termed waste heat performance ratio (PR_{WH}), which encourages the maximal use of the enthalpy of the low-grade heat source relative to the heat sink, is used to facilitate an effective benchmark among low-grade sensible heat-driven processes [1,6].

Stemming from the design of waste heat–driven chillers [7], we posit a modified performance ratio, PR_{WH}, so as to address sensible (waste) heat sources. The actual energy input into the system, which applies to conventional *PR*, is now superseded by the maximal available energy from the heat source. The PR_{WH} can then be defined as:

$$PR_{WH} = \frac{\dot{m}_d \cdot \Delta h_{ref}}{\dot{m}_{hs} \cdot \Delta h_{available}} = \frac{\dot{m}_d \cdot \Delta h_{ref}}{\dot{m}_{hs} \cdot \left(h_{f,hs,in} - h_{f,cond,in}\right)} \tag{6.3}$$

where $\Delta h_{available}$ represents the maximum exploitable energy of the heat source relative to the lowest available temperature, which in this case is the condenser inlet temperature. This assumption is valid if the rejected heating medium cannot be utilized for any other valuable application and the excess heat is purged to the environment. This situation is prevalent in many water-intensive mining and refining operations.

Fig. 6.3 compares the *PR* and the PR_{WH} for an MED process that is coupled with a 75°C heat source inlet temperature. By increasing the number of effects from 3 to 7, the heat

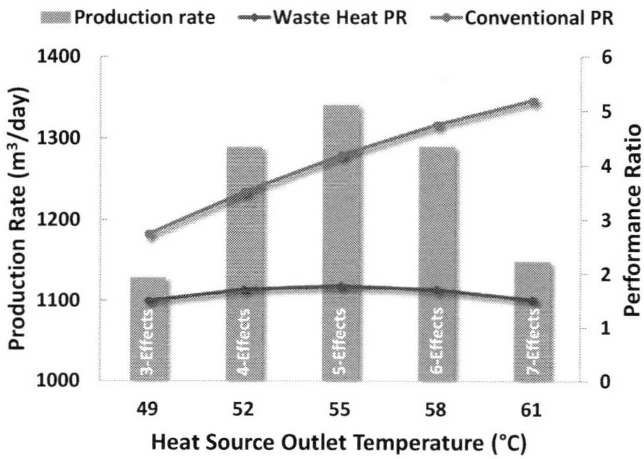

Figure 6.3
Production rate, conventional performance ratio, and waste heat performance ratio for a multi-effect distillation process with 75°C inlet heat source temperature and 100 kg/s mass flow rate scenario for different number of effects.

source outlet temperature decreases; however, an optimal production rate exists. This parameter will maximize at the 5-effect scenario. As Fig. 6.3 shows, the trend of conventional PR is not compatible with the trend of production rate. In contrast the PR_{WH} has the same trend as the production rate. It underscores the incongruence of PR as an efficiency measure for these applications and put in sharp relief the strong correlations between our proposed PR_{WH} and the production rate.

References

[1] A. Christ, K. Regenauer-Lieb, H.T. Chua, Thermodynamic optimisation of multi-effect-distillation driven by sensible heat sources, Desalination 336 (2014) 160–167.

[2] L. Awerbuch, Understanding of Thermal Distillation Desalination Processes, IDA Academy, Singapore, 2012.

[3] T. Pankratz, J. Tonner, Desalination, second ed., Lone Oak Publishing, 2009.

[4] A. Ophir, F. Lokiec, Advanced MED process for most economical sea water desalination, Desalination 182 (1–3) (November 2005) 187–198.

[5] C. Sommariva, Water Management and Economics, IDA Academy, Singapore, 2012.

[6] B. Rahimi, A. Christ, K. Regenauer-Lieb, H.T. Chua, A novel process for low grade heat driven desalination, Desalination 351 (October 2014) 202–212.

[7] H.T. Chua, K.C. Ng, a. Malek, T. Kashiwagi, a. Akisawa, B.B. Saha, Multi-bed regenerative adsorption chiller – improving the utilization of waste heat and reducing the chilled water outlet temperature fluctuation, Int. J. Refrig. 24 (2001) 124–136.

CHAPTER 7

Thermo-Economic Analysis

7.1 Introduction

Many works have reviewed or evaluated the economics of desalination technologies. Hitherto the focus of such works is predominantly a review of the state of the art [1–16]. Less frequent are reports on the economic viability of upcoming technologies that have yet to reach the stage of prototyping. Furthermore, most authors focused on the unit cost of distillate when comparing the economic viability of desalination technologies, with less priority given to annual cash flows, capital investment requirements, and the associated operating costs. According to Badiru and Newnan [17,18], greater importance should be placed upon cash flows and the inherent monetary value of plant installations, implying that capital budgeting metrics such as net present value (*NPV*) and internal rate of return (*IRR*) are of greater use for comparing engineering projects than estimates of unit product cost. This chapter introduces a generalized method that was used to quantitatively compare the economic value of desalination installations using those low-grade sensible heat-driven desalination processes namely multi-effect distillation (MED), preheated (P)-MED, boosted (B)-MED, and flash-boosted (FB)-MED, via such metrics as *NPV* and *IRR* [19].

Capital budgeting metrics such as *NPV*, *IRR*, and incremental rate of return (ΔIRR) can be used to quantitatively assess whether an investment has a greater economic value than an alternative investment. By assuming that all conventional MED, P-MED, B-MED, and FB-MED installations are subject to the same site-specific conditions, the use of these metrics can be extended to assess which process is more economical for the assumed conditions. To produce such metrics, an estimate of the annual cash flows is necessary. Hence estimates of the initial investment, as well as those annual operating expenses, are required.

7.2 Capital Cost Analysis

The initial investment required for a desalination plant is assumed to be equal to the total capital cost of the plant, which is assumed to be inclusive of such costs as engineering, commissioning, erection, equipment, instrumentation, electronics, and control. Reputable data of engineering, procurement, and construction costs are available for conventional MED and multistage flash (MSF) desalination plants from the Global Water Intelligence Desalting Plant Inventory [20], and an estimate of total plant capital cost as a function of

Low Grade Heat Driven Multi-Effect Distillation and Desalination. http://dx.doi.org/10.1016/B978-0-12-805124-5.00007-3

production capacity can be interpolated from such data [19,21]. For this analysis, the International Desalination Association census data [20] are deemed suitable for the purposes of a generalized analysis. A review by Park et al. [9] asserts that location variance has little appreciable effect upon the cumulative costs of a desalination plant, hence all those data were considered regardless of the plant location. The spread of data pertaining to the capital costs of MSF and MED plants indicated that a regression function could be derived for both MED and MSF plants as a function of daily plant capacity in the bounded region shown in Fig. 7.1, and highlights the dense conservative limit of data points [19]. To improve the accuracy of the regression analysis, obvious outliers were removed, leaving only the conservative values within the upper limit of the high density region. The regression equation obtained for the remaining data therefore generates a conservative estimate of capital costs for plants of production capacities up to 10,000 m^3/day—a range that encompasses the capacities of all the simulated cases. Eq. (7.1) obtained via the regression analysis on the adjusted data gives an estimate of the capital cost of both the MED and MSF plants as a function of their respective daily production capacities [19].

$$TCC_{MED,\ MSF}(\text{US\$}) = \Psi_{D_t} = 3054 \times D_t^{0.9751} \tag{7.1}$$

where TCC is the total capital cost of the desalination plant and D_t is the total production rate of a conventional plant in m^3/day. This function ($R^2 = 0.994$) in the bounded region of the relevant data (shown with two dashed lines) shows that the conservative capital cost can be considered as a function of plant production capacity in this region [19,21].

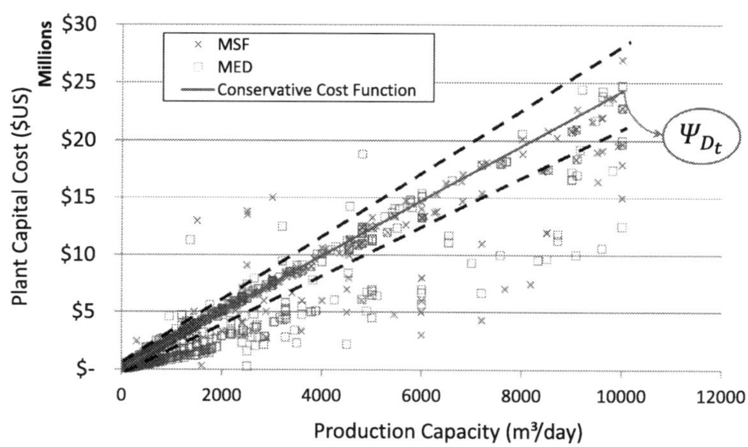

Figure 7.1
Thermal (MED and MSF) desalination plant capital costs as a function of daily production capacity. Dashed lines have been added to bound the region of high density upper limit of data points.

Sommariva et al. [22–24] subdivided the capital costs for MED and MSF systems into five main sectors: the capital cost of the evaporator (40% of the total capital cost); equipment and piping (29%); erection (14%); engineering and commissioning (10%); and electrical, instrumentation, and control (7%).

The total capital cost of a P-MED plant is evaluated by Eq. (7.2) as the sum of the capital cost of the primary MED section (the first term) and the capital cost of the heat exchangers that are used as preheaters (the second term). Based on the types of heat exchanger (plate, shell, tube, etc.) the second term can be calculated (see Appendix E for plate-type heat exchangers).

$$TCC_{P-MED}(\text{US\$}) = \Psi_{D_t} + \sum_{i=1}^{j} CC_{prh,i} \tag{7.2}$$

The total capital cost of a B-MED plant is approximated as the sum of the capital cost of the primary MED section, specified as the first term of Eq. (7.3), augmented for the increased heat transfer requirements by the second term, and the cost of the booster unit is then specified by the third term [19]. The second term contained within brackets is an augmenting factor that accounts for the increased heat transfer requirements of the primary MED effects due to vapor injection. By assuming similar overall heat transfer coefficients for both evaporation and condensation processes [25], only half of the typical heat transfer area to process the additional vapor injection is therefore needed, which is then translated to a 50% discount of the earlier introduced cost breakdown for the evaporator (namely 50% of 40% for the cost of evaporator [22,23], which translates to 20%). This overall factor is then applied to the capital cost differential between an MED plant with a production capacity of a B-MED plant and an MED plant with a production capacity of a B-MED plant less the total vapor injection rate from the booster. The cost of booster unit is approximated as the cost of an MED plant with a daily production capacity equal to the production capacity of the booster unit itself. This cost is then discounted by a factor of 69%, as it is assumed that the cost of the booster unit is inclusive only of evaporator (40% of the total capital cost), equipment and piping costs (29% of the total capital cost) [19,21,24].

$$TCC_{B-MED}(\text{US\$}) = \Psi_{D_{t,B-MED}-D_{Booster}} + \left[0.20\cdot\left(\Psi_{D_{t,B-MED}} - \Psi_{D_{t,B-MED}-D_{Booster}}\right)\right] + 0.69\cdot\Psi_{D_{Booster}} \tag{7.3}$$

For an FB-MED plant, the capital cost is estimated to be equal to the sum of cost of the primary MED section and the flashing chambers, with an augmenting cost factor to account for the increased heat transfer requirements of the primary MED effects due to supplementary steam injection. Eq. (7.4) specifies the cost of FB-MED [19,21]. The first term relates to the primary MED effects of the FB-MED, based upon the vapor production of these effects. The second term contained within brackets is an augmenting factor that accounts for the increased heat transfer requirements of the primary MED effects due to vapor injection. Again as explained before, a 20% factor is then applied to the capital cost differential between an MED plant with a production capacity of an FB-MED plant and an MED plant with a production capacity of an FB-MED plant less the total vapor injection rate from the flashing chambers. The third term specifies the cost of the flashing chambers that augment the vapor production of the system. In contrast to an MSF plant, no condenser tubing and no heat rejection section is required for the flashing chambers, which are major cost items in thermal desalination plants [22–24]. For this purpose and with reference to the cost breakdown of thermal desalination plants approximately 30% of the evaporator costs are associated with the flashing chambers in an MSF plant (namely, 30% of 40%, which is a 12% applied factor) [19]. By considering the cost of equipment and piping as 29% of the total capital cost, the capital cost for the flashing chambers can therefore be estimated with $(0.41 \cdot \Psi_{D_{FC}})$.

The fourth term is the capital cost of the brine heater, which in the case of using a plate-type liquid–liquid heat exchanger can be conservatively evaluated by the relevant equations in Appendix E.

$$TCC_{FB-MED}(\text{US\$}) = \Psi_{D_{t,FB-MED}-D_{FC}} + \left[0.20 \cdot \left(\Psi_{D_{t,FB-MED}} - \Psi_{D_{t,FB-MED}-D_{FC}}\right)\right] + 0.41 \cdot \Psi_{D_{FC}} + CC_{hex} \tag{7.4}$$

7.3 Operating Cost Analysis

Annual operating expenses constitute the remaining costs of a desalination facility. The annual operating expenses were assumed to be comprised of electrical, labor, chemical additives, maintenance and spares, and insurance costs (Eq. (7.5)). Constancy of operating conditions, external economic conditions, and plant reliability is assumed, such that annual operating expenses remain constant over the useful life of a plant, albeit with an assumed escalation factor to account for inflation in the cost of goods and services.

$$OPEX_m = (AEC + LC + ChC + MSIC) \cdot (1 + ES)^{m-1}; \quad m \in \{1, 2, 3, \ldots, n\} \tag{7.5}$$

where $OPEX_m$ is the operating cost in each year m. AEC, LC, ChC, and $MSIC$ are the annual electrical costs; labor costs; chemical additive costs; and maintenance, spares, and insurance costs, respectively, in the first year. ES is the escalation rate, which shows the impact of the inflation rate per year. For the sake of simplicity, by ignoring the escalation rate all these costs can be considered unchanged annually so that $OPEX$ does not change during the lifetime of a plant.

7.3.1 Electricity

Electrical expenses are approximated to the cost of electricity consumed by all process pumps, as these should constitute the main contribution to the total electrical energy consumption of a desalination plant. The rate of consumption of electricity is known for each process pump owing to existing process simulations [21]. The annual electrical cost (*AEC*) is specified by Eq. (7.6) [19].

$$AEC(\text{US\$/yr}) = EUP(\text{US\$/kWh})\cdot p(\text{kWh/m}^3)\cdot D_t(\text{m}^3/\text{day})\cdot f\cdot 365(\text{day/yr}) \qquad (7.6)$$

where the electricity unit price, *EUP*, is a regional parameter and in our case (see Chapter 8) is assumed to have a value of 0.151 AU$/kWh based upon published Australian rates [19,26], which was converted to US$ based upon an exchange rate [27] of 1.086 AU$ per US$ in 2014 (however, this exchange rate varies and should be updated by the time of evaluation). The plant availability, *f*, is assumed to be 95% per year [19,21].

Referring to Chapter 5, our considered processes require five pumps inclusive of the saline water intake pump, heat source pump, noncondensable gas extraction (vacuum) pump, distillate pump, and brine blowdown pump. The improved thermal efficiency and production rates of P-MED and B-MED are enabled by an increased duty requirement of these pumps. For an FB-MED process this is enabled by increased duty on these pumps, as well as by the addition of a brine recirculation pump, make-up water pump, and drain pump, which service the flashing chambers that boost the production of the primary MED effects. All details related to the pumping power calculation including the efficiencies and relevant pressure heads have been explained in detail in Chapter 5.

7.3.2 Labor

Labor costs are based upon actual cost data, and the estimate is made that one full-time equivalent skilled personnel is required per 4543 m^3/day plant capacity [19]. This factor is also a regional parameter and changes in different regions, countries, and locations. The average cost of such a water services sector employee in Australia is 97,905 US$ per year [19,28]. All simulations considered for the purposes of this evaluation have production rates below 4543 m^3/day (see Chapter 8), hence it is assumed that the labor cost (*LC*) is fixed at the prescribed value, as shown by Eq. (7.7) [19].

$$LC(\text{US\$/yr}) = 97905(\text{US\$/yr})\cdot f \qquad (7.7)$$

7.3.3 Chemicals

Chemical additives are necessary for ensuring the effectiveness and maintainability of the desalination process, the potability of product, and constitute a necessary annual expense. The primary MED feed water has specific chemical dosing requirements, thereby constituting the total chemical expense of a conventional MED. Due to the same structure of P-MED and B-MED processes with the conventional MED process, the specific chemical dosing and cost (based on the feed water volume) will be calculated in the same way and is considered to be 0.0223 US\$/m^3 [13,19,25].

The FB-MED has additional chemical requirements due to the required chemical treatment of the flashing chamber feed. The specific chemical dosing cost of the primary MED feed is considered to be 0.0223 US\$/m^3 as mentioned before based upon values from [13,19,25], and the specific chemical cost for the flashing chamber feed is assumed to be 0.0198 US\$/m^3 based upon values from [6,12,13,19,25]; assuming that the chemical requirements of the flashing chamber feed are similar to the feed of MSF plant evaporators at low temperature.

Eq. (7.8) can estimate the annual chemical cost (*ChC*) for the aforementioned sections (primary MED or flashing chambers).

$$ChC(\text{US\$/yr}) = \dot{m}_{F,total}(\text{kg/s}) \cdot CUP\left(\text{US\$/m}^3\right) \cdot 3600(\text{s/hr}) \cdot 24(\text{hr/day}) \times 365(\text{day/yr}) \cdot f/\rho_F\left(\text{kg/m}^3\right) \quad (7.8)$$

where *CUP* is the chemical unit price, which is explained earlier, and $\dot{m}_{F,total}$ is the total feed water flow rate of the considered section.

7.3.4 Maintenance, Spares, and Insurance

The yearly cost of maintenance activities, spares, and insurance (*MSIC*) is assumed to be 1.5% of the total capital cost of the plant [29], as specified by Eq. (7.9) [19]. The cost of maintenance and spares of any desalination plant are highly dependent upon factors such as the timing of cash flows as well as the proficiency of engineering services and management. Therefore the annual costs of a plant utilizing the well-known MED process could differ greatly to the cost of similar plants utilizing the P-MED, B-MED, or FB-MED process, especially in the infancy of these technologies. However, without any justifiable alternative, such costs are approximated as a percentage of the total capital cost of the plant [19].

$$MSIC(\text{US\$/yr}) = 1.5\% \times TCC \quad (7.9)$$

where the total capital cost, *TCC*, is determined as a function of production capacity in the manner described in Section 7.2.

7.4 Cash Flow and Capital Budgeting Metrics

Annual operating expenses, income taxes, loan repayments, and incomes are assumed to comprise the annual cash flows over the lifetime of the desalination plants, with all incomes and expenses assumed to increase at a rate of 3% per year [19] from the starting year to account for inflation (however, this rate changes for different countries and locations). Cash flows are then used to determine capital budgeting metrics so that the processes can be compared. It is assumed that the initial investment is equal to the total capital cost of the desalination plant, and this cost is incurred exhaustively at the beginning of the plant's useful life.

7.4.1 Incomes

It is assumed that the incomes of the desalination plant in a period are equal to the incomes made on the sale of the volume of water equal to the total production of the plant within that same period, sold at the average market price [19]. Therefore, at the end of each year the income can be calculated by:

$$Income_m = 365 \cdot D_t\left(\mathrm{m}^3/\mathrm{day}\right) \cdot f \cdot WMP\left(\mathrm{US\$}/\mathrm{m}^3\right) \cdot (1 + ES)^{m-1} \tag{7.10}$$

where $Income_m$ is the income at the end each year (m is between 1 and n); WMP is the market price of water product (distillate) available from an external supplier, assumed as being equal to the Australian average water price of 2.72 AU\$/m^3 [30], which is then converted into US$ based upon current exchange rates [27]; and ES is the escalation rate, which is different in different regions and countries. In our simulation a 3% escalation rate is applied to calculate the income for the following years [19]. It should be noted that all these assumptions are dependent upon regions, countries, and time.

7.4.2 Costs

The major costs include operating cost, depreciation, loan repayment, and income tax. The operating cost has been explained in Section 7.3. A simple straight line (SL) depreciation is assumed to apply to the capital cost of a plant, Eq. (7.11).

$$SL_m = \frac{CA - SV}{n} \tag{7.11}$$

where CA is the cost of the asset (which in our case can be considered as TCC); SV is the salvage value, which in our case is considered to be zero; and n is the total plant life, which is taken to be the depreciable life. The calculated SL by Eq. (7.11) is a uniform annual depreciation cost.

For the loan repayment, it is assumed that a bank loan covers 50% (BLP) of the cost of the initial investment, equal to the total capital cost of the plant, with a 2% supplementary fee (S) to cover miscellaneous bank fees, and with interest incurred annually at a rate of 8% upon borrowed money [19] (these assumptions may and can change in different regions, countries, locations, and bankers). Annual fixed loan repayments, Eq. (7.12), are made to cover the cost of loan repayment, assuming the loan is to be fully paid off by the end of the plant's useful life, which is assumed to be 30 years for all cases. Eq. (7.13) specifies the capital recovery factor, *CRF*, a factor that enables the calculation of a single amortizing value for an amount that is to be fully amortized within n years, and which incurs interest at an annual rate, i [12,18,25].

$$PMT_m = (1 + S) \cdot BLP \cdot TCC \cdot CRF \tag{7.12}$$

$$CRF = \frac{i \cdot (1 + i)^n}{(1 + i)^n - 1} \tag{7.13}$$

PMT also has a uniform annual value over the plant lifetime.

The income tax, Eq. (7.14), is calculated by assuming that the tax rate is a flat 30% of taxable income [31] (that in this case is the summation of income, operating cost, depreciation, and annual bank interest payment); however, income tax will often affect the alternatives similarly, allowing to compare the choices without considering income taxes. Tax rate also changes in different countries and locations. The total cash flow in a year is then equal to the difference of the total annual benefits and total annual expenditures, adjusted for inflation and for tax.

$$ITAX_m = TR \cdot (Income_m + OPEX_m + SL_m + IPMT_m) \tag{7.14}$$

where $ITAX_m$ is the annual income tax in each year and *TR* is the tax rate. $IPMT_m$ is the annual bank interest payment, which is calculated based on the uniform annual repayment and the loan amount. Table 7.1 shows the method to find the principal ($PPMT_m$) and interest payment ($IPMT_m$) of a loan repayment.

In Table 7.1, LA_0 is the total loan amount borrowed from the bank in the zeroth year. LA_n is the remaining loan balance at the end of the due time, which should be equal to zero. The interest and principal payments are the portions of uniform annual loan repayment in each year so that $PMT = PPMT_m + IPMT_m$. Excel spreadsheets also have some functions

Table 7.1: Relevant equations for principal and interest payments.

Year	Ending Balance	Uniform Annual Loan Repayment	Interest Payment	Principal Payment
0	$LA_0 = (1 + S) \cdot BL \cdot TCC$	–	–	–
1	$LA_1 = LA_0 - PPMT_1$	$PMT_1 = LA_0 \cdot CRF$	$IPMT_1 = i \cdot LA_0$	$PPMT_1 = PMT - IPMT_1$
2	$LA_2 = LA_1 - PPMT_2$	$PMT_2 = LA_0 \cdot CRF$	$IPMT_2 = i \cdot LA_1$	$PPMT_2 = PMT - IPMT_2$
3	$LA_3 = LA_2 - PPMT_3$	$PMT_3 = LA_0 \cdot CRF$	$IPMT_3 = i \cdot LA_2$	$PPMT_3 = PMT - IPMT_3$
4	$LA_4 = LA_3 - PPMT_4$	$PMT_4 = LA_0 \cdot CRF$	$IPMT_4 = i \cdot LA_3$	$PPMT_4 = PMT - IPMT_4$
...	...	...	...	...
$n-1$	$LA_{n-1} = LA_{n-2} - PPMT_{n-1}$	$PMT_{n-1} = LA_0 \cdot CRF$	$IPMT_{n-1} = i \cdot LA_{n-2}$	$PPMT_{n-1} = PMT - IPMT_{n-1}$
n	$LA_n = LA_{n-1} - PPMT_n = 0$	$PMT_n = LA_0 \cdot CRF$	$IPMT_n = i \cdot LA_{n-1}$	$PPMT_n = PMT - IPMT_n$

to calculate *PMT*, *PPMT*, and *IPMT*. These functions have the same name as what we have used in this chapter (*PMT*, for finding the uniform annual payment; *PPMT*, for the principal payment; and *IPMT* for the interest payment).

After finding $ITAX_m$ from Eq. (7.14), the after tax income can be calculated by

$$ATAX\ Income_m = Income_m + OPEX_m + PMT_m + ITAX_m \tag{7.15}$$

$ATAX\ Income_m$ is the net benefit in each year that the investor earns from operating the plant.

7.4.3 Net Present Value and Choosing the Best Alternative

With cash flows specified over the lifetime of each plant design, the economic metrics can then be calculated. These metrics, often referred to as capital budgeting metrics, allow investments and projects to be compared for their economic value. The *NPV* is widely considered to be the superior and most appropriate metric for ranking the economic value of investments and projects [17,18,32], and it is taken to be the primary means of comparison in this analysis. *NPV*, as specified by Eq. (7.16) [18], is the sum of the annual cash flows of a project or investment, discounted to a present day value by means of an interest rate, *i*.

$$NPV = \text{Present Value of Benefits} - \text{Present Value of Costs} = \sum_{m=1}^{n} \frac{ATAX\ Income_m}{(1+i)^m} - TCC \tag{7.16}$$

The *IRR* is the interest rate at which the *NPV* is equal to zero, where the benefits are equivalent to the costs. To compare two or more alternatives the ΔIRR metric and incremental analysis can be the preferred method [17,18]. An incremental analysis compares the cash flows of two specific investments, or engineering projects, over the lifetime of the projects. The ΔIRR is the value of the interest rate, which yields an *NPV* of zero for the differential cash flows of two projects [18].

The minimum acceptable rate of return (*MARR*) is assumed to be equal to the interest rate on borrowed money of 8% (this value changes in different countries and bankers), in projects accompanied by normal business risk and uncertainty [18,19]. Unless the *MARR* is known, neither *NPV* nor annual cash flow analysis is possible, therefore this parameter is an essential parameter for the aforementioned analyses.

For a better understanding, Fig. 7.2 shows the *NPV* curves for three exemplary scenarios (*A*, *B*, and *C*) that have been plotted against interest rate. As Eq. (7.17) shows, by knowing the after-tax income in each year and the *TCC*, the *NPV* can be plotted based on the interest rate (Fig. 7.2). In order to choose the best scenario, we need to analyze this graph

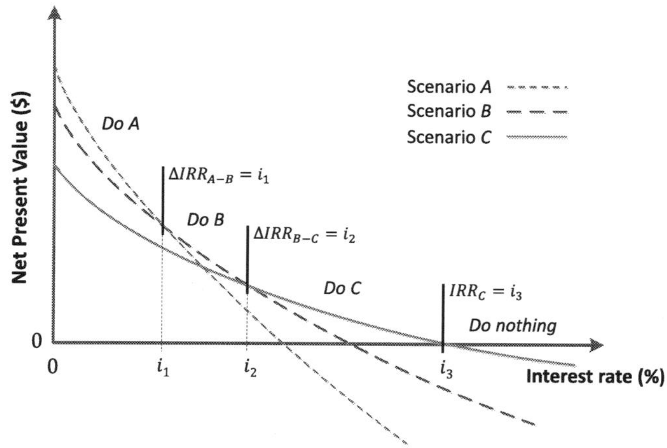

Figure 7.2
Net present value graphs for three different scenarios.

for different interest rates. As it shows for an interest rate between 0 and i_1, scenario *A* has the highest *NPV* among the various scenarios, therefore this scenario is the best choice for any *MARR* within this range. The same methodology is applied to analyze for the other ranges of interest rate so that the scenario that yields the highest *NPV* is the best choice within a specific range of interest rate. The most interesting issue is that the do-nothing scenario can sometimes be the best choice. As Fig. 7.2 shows, the *NPV* of scenario *C* will be zero at i_3. In this case for any *MARR* more than i_3, the do-nothing scenario is the rational. Eq. (7.17) summarizes Fig. 7.2.

$$Best\ Scenario = \begin{cases} A; & 0 \le MARR \le i_1 \\ B; & i_1 \le MARR \le i_2 \\ C; & i_2 \le MARR \le i_3 \\ Do\text{-}Nothing; & i_3 \le MARR \le \infty \end{cases} \tag{7.17}$$

In the case of ΔIRR analysis, as mentioned earlier, for example for two scenarios *B* and *C*, ΔIRR_{B-C} (namely i_2) shows the interest rate at which the *NPV* of these two scenarios are equal. It means that this point is the threshold for a change in decision making in terms of selecting the best scenario.

7.4.4 Unit Product Cost

The unit product cost (*UPC*) of water, Eq. (7.18), is also used as an appropriate metric for comparing desalination plants and processes [5,8,10,12,25,33]. The use of the *UPC* metric is appropriate for ranking the production costs of desalination plants; however, the plant with the lowest specific production costs is not inherently the plant that will be the most profitable. Taxation, asset depreciation, interest charges, among others, also affect the profitability of an engineering project. A desalination plant is a monetary investment, as well as an engineering endeavor, and therefore it must be framed and compared to alternatives in terms of profitability and worth. Pasqual [32] asserted that the consensus on ranking monetary investments was that metrics such as *NPV* were the best means of ranking investment alternatives.

$$UPC = \frac{TCC \times CRF + OPEX}{365 \cdot f \cdot D_t} \tag{7.18}$$

For the sake of simplicity, in Eq. (7.18) the impact of escalation factor in the operating cost has been ignored, therefore *OPEX* refers to the operating expense incurred in the first year of operation and a uniform annual *UPC* can be preliminarily estimated.

In order to factor in the impact of escalation factor in the operating expenses during the lifetime of the plant, the *UPC* can be calculated based on an *NPV* analysis. In this method, the market price of water (see Section 7.4.1), which makes the *NPV* equal to zero (as in Eq. (7.16)), can be considered to be the uniform *UPC* during the lifetime of the plant. This method of evaluating the *UPC* will be used in the subsequent chapters.

References

[1] I.C. Karagiannis, P.G. Soldatos, Water desalination cost literature: review and assessment, Desalination 223 (2008) 448–456.
[2] R. Borsani, S. Rebagliati, Fundamentals and costing of MSF desalination plants and comparison with other technologies, Desalination 182 (1–3) (November 2005) 29–37.
[3] A.M. El-Nashar, Why use renewable energy for desalination, in: Desalination and Water Resources (DESWARE), Renewable Energy Systems and Desalination, vol. 1, EOLSS, 2010, pp. 202–215.
[4] N.M. Wade, Distillation plant development and cost update, Desalination 136 (2001) 3–12.
[5] W. El-Mudir, M. El-Bousiffi, S. Al-Hengari, Performance evaluation of a small size TVC desalination plant, Desalination 165 (2004) 269–279.
[6] H.M. Ettouney, H.T. El-Dessouky, R.S. Faibish, P.J. Gowin, Evaluation the economics of desalination, Chem. Eng. Prog. (CEP) (December 2002) 32–39.
[7] P. Fiorini, E. Sciubba, Thermoeconomic analysis of a MSF desalination plant, Desalination 182 (1–3) (November 2005) 39–51.
[8] N. Lior, Advances in Water Desalination, John Wiley & Sons, Inc., 2013.
[9] M. Park, N. Park, H. Park, H. Shin, B. Kim, An economic analysis of desalination for potential application in Korea, Desalination 114 (1997) 209–221.
[10] O.J. Morin, Design and operating comparison of MSF and MED systems, Desalination 93 (1993) 69–109.
[11] O.J. Morin, Process optimization – cost criteria, in: Desalination and Water Resources (DESWARE), Thermal Desalination Process, vol. 1, EOLSS, 2010, pp. 188–209.
[12] A.S. Nafey, H.E.S. Fath, A.A. Mabrouk, Exergy and thermoeconomic evaluation of MSF process using a new visual package, Desalination 201 (1–3) (November 2006) 224–240.
[13] A.S. Nafey, H.E.S. Fath, A.A. Mabrouk, Thermo-economic investigation of multi effect evaporation (MEE) and hybrid multi effect evaporation—multi stage flash (MEE-MSF) systems, Desalination 201 (1–3) (November 2006) 241–254.

[14] A.S. Nafey, H.E.S. Fath, A.A. Mabrouk, Thermoeconomic design of a multi-effect evaporation mechanical vapor compression (MEE−MVC) desalination process, Desalination 230 (1−3) (September 2008) 1−15.
[15] A.M. El-Nashar, The economic feasibility of small solar MED seawater desalination plants for remote arid areas, Desalination 134 (2001) 173−186.
[16] K. Thu, A. Chakraborty, B.B. Saha, W.G. Chun, K.C. Ng, Life-cycle cost analysis of adsorption cycles for desalination, Desalin. Water Treat. 20 (2010) 1−10.
[17] A.B. Badiru, O.A. Omitaomu, Computational Economic Analysis for Engineering and Industry, Taylor and Francis, Hoboken, 2007.
[18] D.G. Newnan, J.P. Lavelle, T.G. Eschenbach, Engineering Economic Analysis, Oxford University Press, 2009.
[19] B. Rahimi, J. May, A. Christ, K. Regenauer-Lieb, H.T. Chua, Thermo-economic analysis of two novel low grade sensible heat driven desalination processes, Desalination 365 (2015) 316−328.
[20] IDA Desalting Plants Inventory, 2011 (Online). Available: http://www.desaldata.com.
[21] B. Rahimi, A. Christ, K. Regenauer-Lieb, H.T. Chua, A novel process for low grade heat driven desalination, Desalination 351 (October 2014) 202−212.
[22] C. Sommariva, H. Hogg, K. Callister, Maximum economic design life for desalination plant: the role of auxiliary equipment materials selection and specification in plant reliability, Desalination 153 (2002) 199−205.
[23] C. Sommariva, H. Hogg, K. Callister, Cost reduction and design lifetime increase in thermal desalination plants: thermodynamic and corrosion resistance combined analysis for heat exchange tubes material selection, Desalination 158 (May 2003) 17−21.
[24] C. Sommariva, D. Pinciroli, E. Tolle, R. Adinolfi, Optimization of material selection for evaporative desalination plants in order to achieve the highest cost-benefit ratio, Desalination 124 (1999) 99−103.
[25] H.T. El-Dessouky, H.M. Ettouney, Fundamentals of Salt Water Desalination, Elsevier Science B.V., 2002.
[26] Synergy, Standard Electricity Prices and Charges − South West Interconnected System, 2013 (Online). Available: http://www.synergy.net.au/docs/Standard_Electricity_Prices_Charges_brochure_2014.PDF.
[27] Reserve Bank of Australia, Exchange Rates, Reserve Bank of Australia, 2014 (Online). Available: www.rba.gov.au/statistics/frequency/exchange-rates.html.
[28] Australian Bureau of Statistics (ABS), Labour Costs Australia 2010−2011, 2012.
[29] O.J. Morin, Cost aspects − MSF, in: Desalination and Water Resources (DESWARE), Thermal Desalination Processes, vol. II, EOLSS, 2010, pp. 1−30.
[30] Australian Bureau of Statistics (ABS), Western Australia. State Fact Sheet 2013, 2014 (Online). Available: http://www.abs.gov.au/AUSSTATS/.
[31] ATO, Tax Rates 2012−13, Australian Taxation Office (ATO), 2014 (Online). Available: https://www.ato.gov.au/Rates/Company-tax/.
[32] J. Pasqual, E. Padilla, E. Jadotte, Technical note: equivalence of different profitability criteria with the net present value, Int. J. Prod. Econ. 142 (1) (March 2013) 205−210.
[33] M.K. Wittholz, B.K. O'Neill, C.B. Colby, D. Lewis, Estimating the cost of desalination plants using a cost database, Desalination 229 (1−3) (September 2008) 10−20.

CHAPTER 8

Application of Novel Low Grade Heat-Driven Distillation to Seawater Desalination

8.1 Introduction

The simulation methods for all the configurations namely, conventional multi-effect distillation (MED), preheated MED (P-MED), boosted MED (B-MED), and flash-boosted MED (FB-MED) processes have been explained in detail in Chapter 4. The salient features for all the required pumps have been introduced in Chapter 5, including the relevant pumping power in each process. A cash flow analysis to compare all these configurations has been explained in detail in Chapter 7. Following all aforementioned exposition, a detailed thermoeconomic evaluation for conventional MED, B-MED, and FB-MED processes will be discussed in this chapter. Based on our results, in terms of low grade sensible heat applications, the B-MED process is economically superior to the feed P-MED configuration [1], and hence the latter is not simulated in this chapter.

Process simulations and validation of the considered processes are produced based upon steady-state analysis methods used in combination with appropriate boundary conditions (Table 8.1) [2–4].

Table 8.1 delineates all the assumptions inherent in the present simulation. The heat source medium is water and its flow rate has been fixed at 100 kg/s in all simulations. The inlet seawater temperature at the condenser is 28°C and the salinity of the seawater is 35,000 ppm. Due to the scaling issue, which is a big problem in the first effect of MED plants, the maximum top brine temperature is considered to be 70°C. For the liquid-to-liquid heat exchanger, which here has been considered as a plate-type heat exchanger, the minimum temperature of approach is 3°C. According to the relevant suggestion by the plate heat exchanger manufacturers, a maximum *NTU* equal to 5.0 is considered for each plate heat exchanger [5]. Therefore, in case of higher required *NTU*, two or more plate heat exchangers are considered in series. For vapor injection to happen from the booster (in the B-MED process) or flashing chamber (in the FB-MED process) to the judicious effect in the primary MED plant, a 500 Pa pressure difference is considered to be sufficient [6].

Low Grade Heat Driven Multi-Effect Distillation and Desalination. http://dx.doi.org/10.1016/B978-0-12-805124-5.00008-5

Table 8.1: Assumptions of thermodynamic simulations.

Maximum top brine temperature (°C)	70
Heat source flow rate, $\dot{m}_{HS}$ (kg/s)	100
Heat source temperature (°C)	65, 70, 75, 80, 85, 90
Feed-to-vapor ratio (35% evaporation)	2.857
$T_{cond,in}$ (°C), $T_{cond,out}$ (°C)	28, 38
ΔP_{inj}[a] (Pa)	500
$BPE + NEA$ in each flashing chamber (°C)[b]	1
Liquid-to-liquid heat exchanger approach temperature (°C)[b]	3.0
Minimum temperature drop for the heat source medium in each evaporator (°C)	6.0
Feed salinity, X_F (ppm)	35,000
Maximum NTU for each plate heat exchanger[b]	5.0
$X_{B,FC}$ (ppm), maximum outlet salinity for the flashing section[b]	70,000

[a]Relevant pressure difference for steam injection purpose in boosted multi-effect distillation and flash-boosted multi-effect distillation processes.
[b]Applicable for flash-boosted multi-effect distillation configuration.

The results of the process simulation and thermoeconomic evaluation are presented in the following sections.

8.2 Simulation Results

All results including the production rate, number of effects, pumping power consumptions, and waste heat performance ratio for a range of sensible heat source temperature between 65°C and 90°C are shown in Table 8.2. Figs. 8.1−8.3 show the schematic layout of optimized MED, B-MED, and FB-MED systems, which are coupled to a 65°C sensible heat source. Table 8.3 also includes the main flow rates of the considered processes.

As shown in Table 8.2, the performance of the FB-MED scheme with those of the optimized conventional and B-MED processes have been compared. The number of effects for the three schemes as in Table 8.2 (optimized conventional MED, B-MED, and FB-MED) generally increases with temperature, so as to maximize freshwater production. Subject to the same heat source inlet temperature, the heat source temperature drop across the primary MED plant, for the three schemes presently considered, is generally different so that each technology maximizes its respective freshwater production. Specifically, the heat source temperature drop across the primary plant is generally the smallest for the FB-MED scheme, as the remaining heat source enthalpy can be better exploited at the flashing chambers. This is then followed by the B-MED process, whereby the steam booster unit can still further process the remaining waste heat. The largest heat source temperature drop is verily found across the optimized conventional MED plant, as this is the only avenue to utilize the waste heat. While more effects can certainly be built into the conventional MED system at any particular heat source inlet temperature, as delineated in

Table 8.2: Production rates and power consumption of the conventional, boosted, and flash-boosted multi-effect distillation (MED) processes.

Heat source inlet temperatures (°C)	65	70	75	80	85	90
Optimized Conventional MED						
Freshwater yield (m^3/day)	707	998	1341	1694	2103	2510
Normalized pumping power (kWh/m^3)	2.15	2.10	1.69	1.64	1.38	1.36
Number of effects	4	4	5	5	6	6
Heat source temperature drop (°C)	12.9	18.2	20.5	25.8	28.2	33.7
PR_{WH}	1.18	1.47	1.77	2.03	2.30	2.53
Optimized Boosted MED						
Freshwater yield (m^3/day)	865	1217	1603	2034	2462	2920
Normalized pumping power (kWh/m^3)	2.46	1.88	1.72	1.56	1.47	1.36
Number of MED effects/ Injected effect's number	5/5	6/5	6/5	7/6	7/6	8/7
Production rate ratio[a] (%)	22	22	20	20	17	16
Normalized power consumption ratio[a] (%)	14	−11	1	−5	6	0
Heat source temperature drop (°C)	17.6	19.7	24.7	29.8	34.8	39.9
PR_{WH}	1.44	1.80	2.12	2.43	2.69	2.94
PR_{WH} ratio[a] (%)	22	22	20	20	17	16
Flash-Boosted MED[b]						
Freshwater yield (m^3/day)	945	1417	1922	2426	2928	3430
Normalized pumping power (kWh/m^3)	3.20	2.77	2.54	2.32	2.20	2.13
Number of MED effects/ Flashing stages	5/4	7/6	9/8	10/9	10/9	10/9
Production rate ratio[a] (%)	34	42	43	43	39	37
Normalized power consumption ratio[a] (%)	49	32	50	41	60	57
Heat source temperature drop (°C)	19.5	24.6	29.8	34.9	40.0	45.1
PR_{WH}	1.58	2.09	2.54	2.90	3.20	3.45
PR_{WH} ratio[a] (%)	34	42	43	43	39	36

[a]Relative to the optimized conventional MED alternative.
[b]This is not an optimized flash-boosted MED (FB-MED) configuration. The results are based on the maximum amount of flashing chambers in each scenario. The optimized FB-MED configuration for each scenario is explained in Section 8.3.4.

Table 8.2, this is done at the expense of freshwater production, as less heat via a temperature drop can be exploited [6]. Consequently both the B-MED and FB-MED systems can realize a successively larger number of effects than the conventional MED system for the same heat source inlet temperature.

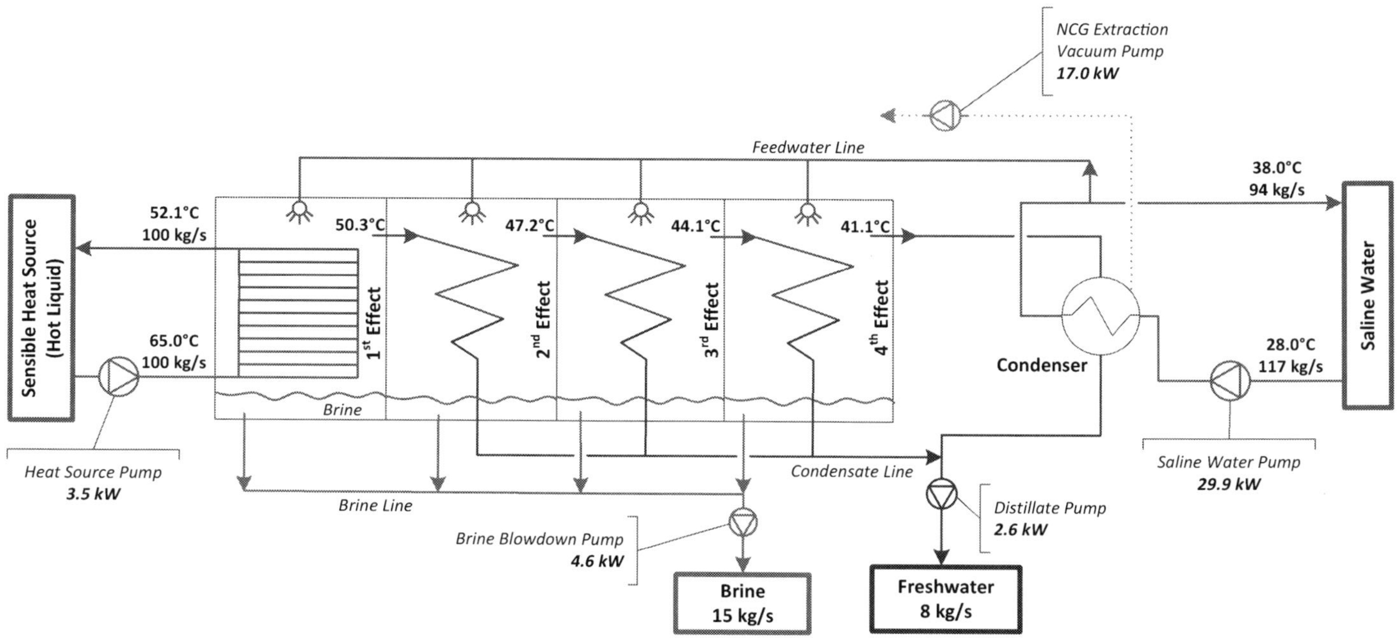

Figure 8.1
Schematic design of an optimized conventional multi-effect distillation system (typical quantities as per the simulation for 65°C inlet heat source temperature).

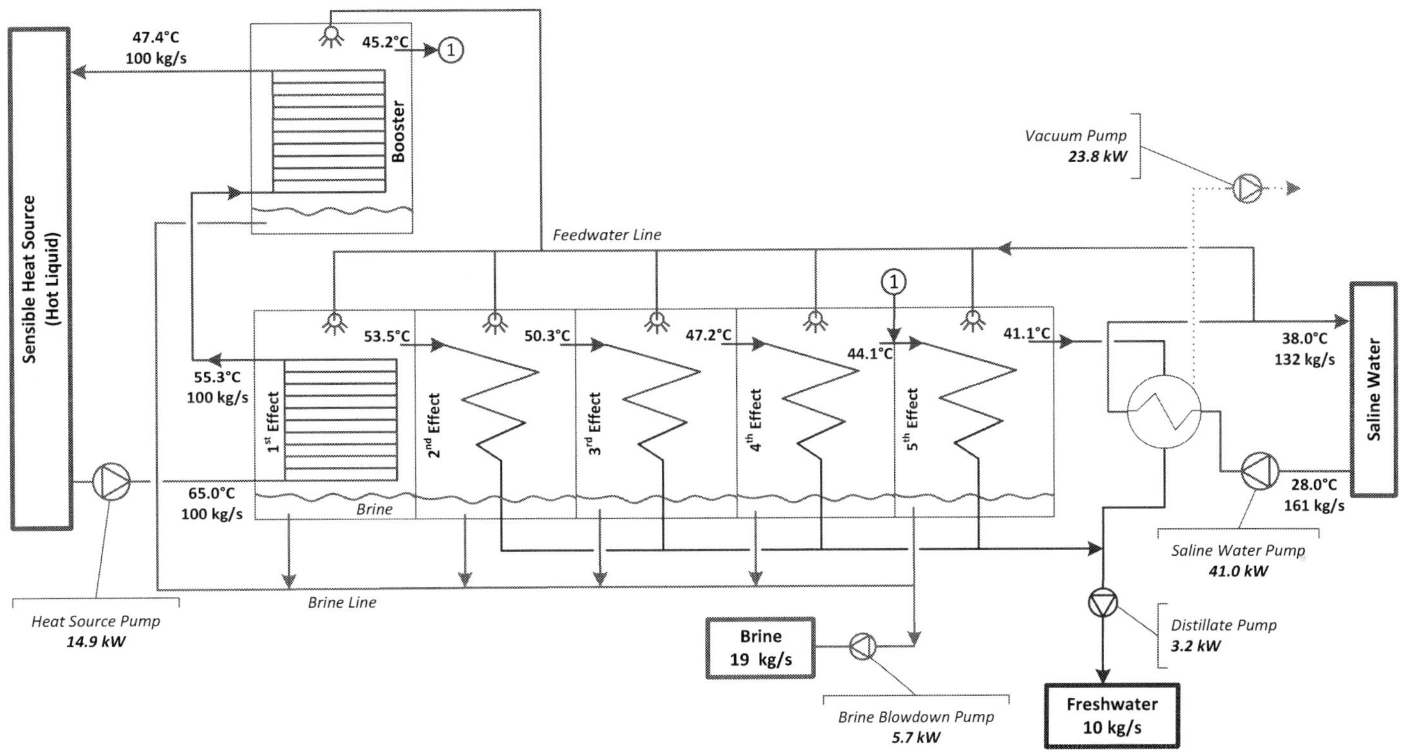

Figure 8.2
Schematic design of an optimized boosted multi-effect distillation system (typical quantities as per the simulation for 65°C inlet heat source temperature).

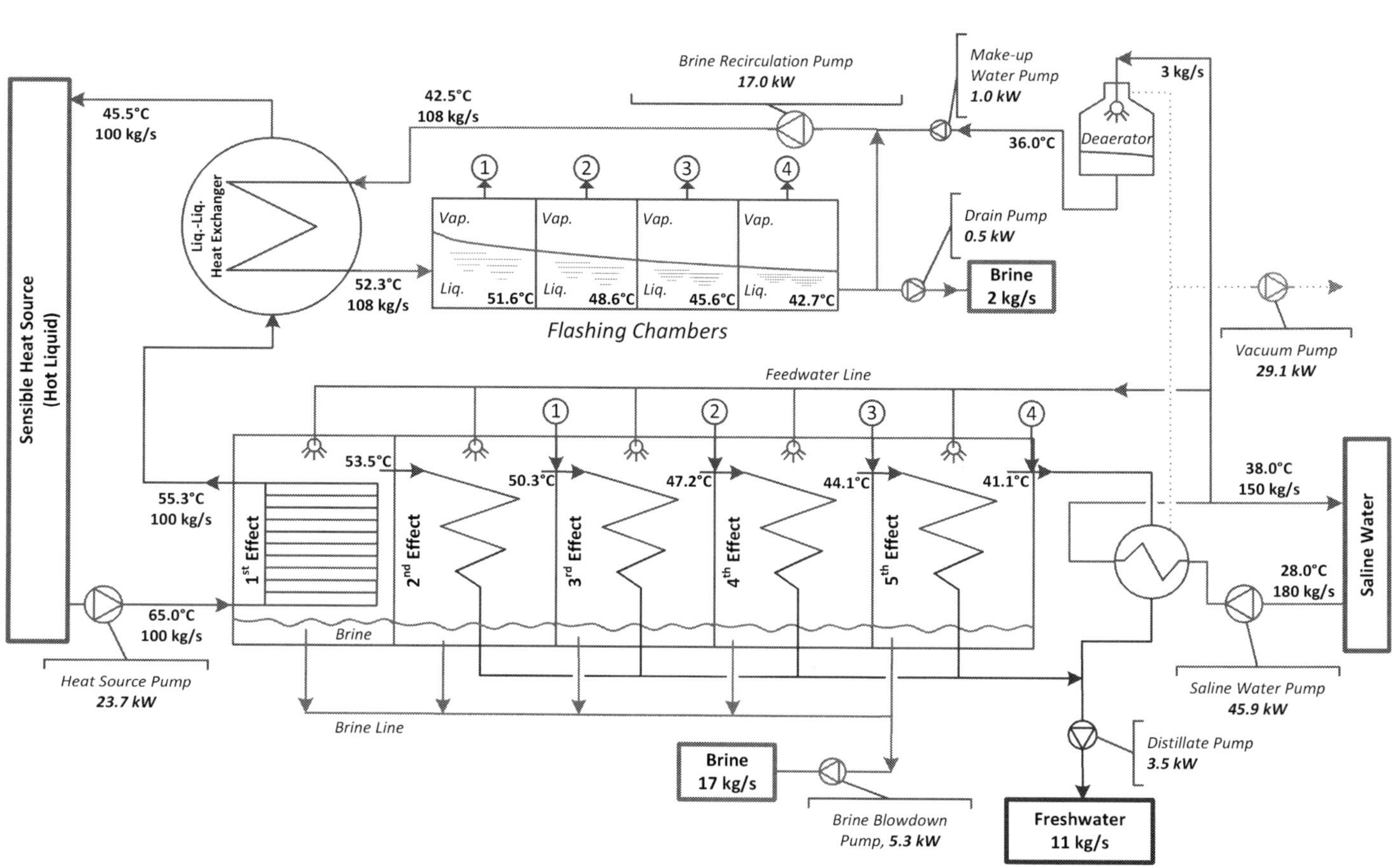

Figure 8.3

Schematic design of an optimized flash-boosted multi-effect distillation system, (typical quantities as per the simulation for 65°C inlet heat source temperature).

Table 8.3: Main flow rates for MED, B-MED, and FB-MED over the considered heat source temperature range.

Type	Flow Rates (kg/s)	Heat Source Inlet Temperature (°C)					
		65	70	75	80	85	90
MED	Condenser inlet	117.1	165.3	173.4	219.0	219.9	262.5
	Condenser outlet	93.9	132.6	129.5	163.6	151.2	180.4
	Freshwater	8.1	11.4	15.4	19.4	24.1	28.7
	Brine	15.0	21.2	28.5	36.0	44.7	53.3
B-MED	Condenser inlet	160.7	167.4	205.4	236.1	269.8	292.1
	Condenser outlet	132.4	127.6	152.9	169.6	189.4	196.7
	Freshwater	9.9	13.9	18.4	23.3	28.2	33.4
	Brine	18.4	25.9	34.1	43.2	52.3	62.0
FB-MED	Condenser inlet	179.9	213.6	241.9	267.9	297.6	327.3
	Condenser outlet	150.5	169.6	182.4	192.7	206.8	220.9
	Freshwater	10.8	16.2	22.0	27.7	33.5	39.2
	Brine	18.7	27.8	37.5	47.5	57.3	67.2

B-MED, boosted multi-effect distillation; *FB-MED*, flash-boosted multi-effect distillation; *MED*, multi-effect distillation.

For clarity, in Fig. 8.1 for the conventional MED plant, the important process temperatures stemming from the simulation are included. The same is done to both the B-MED scheme as in Fig. 8.2, and the FB-MED scheme as in Fig. 8.3. As shown in these figures, at 65°C heat source inlet temperature, each of the MED, B-MED, and FB-MED processes generally requires a different number of effects to maximize the production rate [2,3,6–8]. As Fig. 8.1 shows, the optimized number of effects (based on the relevant assumptions in Table 8.1) for the conventional MED process is four. In Fig. 8.2, the optimized B-MED configuration is able to accommodate one more effect as compared to the optimized conventional MED process on account of a lower temperature drop at the first effect of the primary MED section. This is made possible because of the booster unit, which can process the heat source further, so as to realize an increase to the total production rate by around 22% as compared to the conventional MED [2,7]. In Fig. 8.3, the same reasoning applies to the FB-MED, namely, the higher first effect's heat source outlet temperature of the FB-MED allows it to maintain more effects as compared to both the conventional MED and B-MED processes. The liquid–liquid heat exchanger can then exploit more energy from the heat source and hence the outlet heat source temperature of the FB-MED is around 7°C and 2°C less than that of the conventional MED and B-MED processes, respectively. This exploited energy is transferred to the flashing chamber feed stream and gradually released over the train of flashing chambers and injected into the proper effects of the primary MED section of the FB-MED system, thereby increasing the freshwater generation by around 34% and 10% as compared to the conventional MED and B-MED, respectively. The same methodology [2,3,6–8] is also applied to the other heat source inlet temperatures.

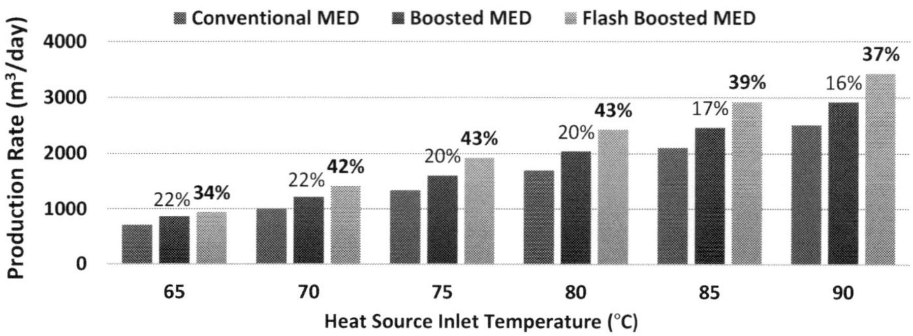

Figure 8.4
Freshwater production rate comparison among the three systems for an assortment of heat source temperatures. The percentages on top of both the boosted and flash-boosted multi-effect distillation columns represent the increment of freshwater production over and above the optimized conventional multi-effect distillation process.

Fig. 8.4 highlights the characteristics of the FB-MED scheme in relation to the conventional and steam-boosted MED processes. Over the range of temperatures from 65°C to 90°C, the FB-MED scheme is 10–20% better in terms of freshwater production than the B-MED process. This potential is better discerned where the production increment rates of both the FB-MED and B-MED schemes are compared against the conventional MED system. The production rate increment decreases gradually for the B-MED process from around 22% at 65°C heat source inlet temperature to close to 16% at 90°C. For the FB-MED scheme, on the other hand, the production rate increment begins at around 34% at 65°C and peaks at about 43% at 75°C, and then pares back to around 37% at 90°C. The reduction is enjoined by the top brine temperature constraint of 70°C at the first effect of the primary MED plant and forces the number of primary MED effects to remain the same for heat source temperatures higher than 80°C [2]. So in general, the freshwater production rate of the FB-MED process over the entire heat source temperature range is more than both the optimized conventional and boosted MEDs. The B-MED enjoys its highest thermal performance in the lower temperature range while the FB-MED performance first increases to 80°C heat source temperature and then slightly decreases; however over the entire temperature range it remains the superior technology.

Fig. 8.5 shows the waste heat performance ratio (see Chapter 6) for all three systems. From this figure, the waste heat performance ratio of the FB-MED experiences a substantial jump at 75°C. This mirrors the boost in production rate at 75°C as shown in Fig. 8.4 [2].

To see the impact of each flashing chamber in each scenario (namely, different heat source inlet temperature) Fig. 8.6 shows the production rate as compared to the optimized conventional MED process, which is shown with "zero" number of flashing chamber. As it shows by progressively adding flashing chambers in each scenario, the production rate is

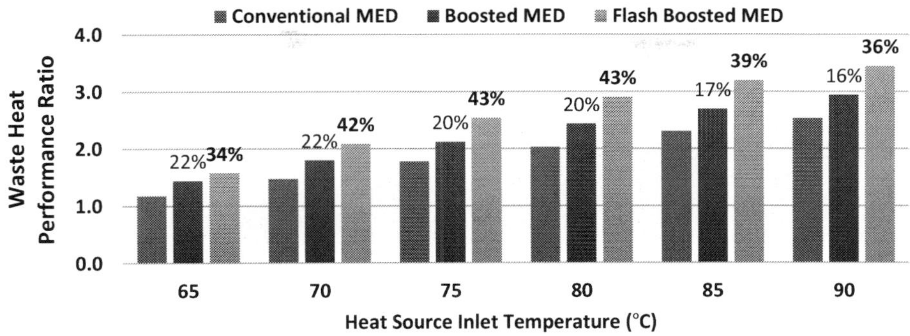

Figure 8.5

Waste heat performance comparison among the three systems and the percentage of increment of boosted and flash-boosted multi-effect distillations' performance ratios compared to the optimized conventional multi-effect distillation for an assortment of heat source temperatures.

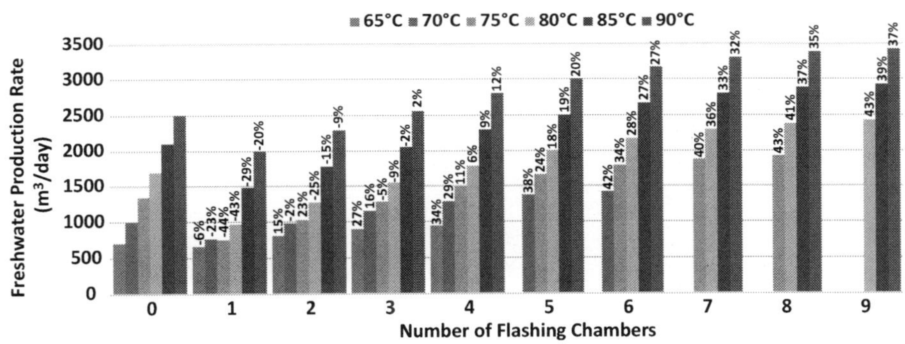

Figure 8.6

Production rate (m^3/day) versus number of installed flashing chambers for all scenarios. The plant designs utilizing zero flashing chambers are optimized conventional multi-effect distillation (MED) plants, while all remaining plant designs are flash-boosted MED (FB-MED). The percentage values beside the bars indicate the percentage increase in the freshwater yield for that FB-MED design compared with the optimized conventional MED alternative (0 number of flashing chamber) for that particular scenario.

increased accordingly; however, it does not necessarily mean that the case with a higher number of flashing chambers is a better option. This issue will be discussed further in Section 8.3.4.

Fig. 8.7 presents the temperature-energy profiles of the heat source medium of all three processes for 75°C heat source inlet temperature scenario. For the conventional MED, the heat source outlet temperature is around 4°C higher than that of the B-MED system and 9°C higher than that of the FB-MED system, although the temperature drop over the first effect is higher than both the B-MED and FB-MED systems. In all three systems, the heat source temperature decreases due to the sensible heat transfer. However for the feed, the

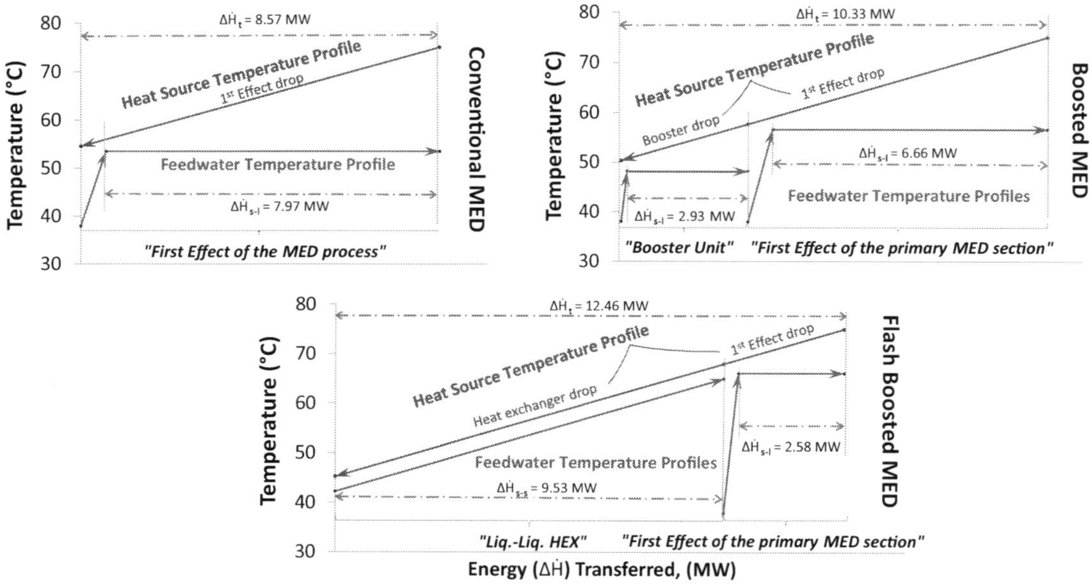

Figure 8.7
Temperature profile for the first effect of the conventional multi-effect distillation (MED), the booster and the first effect of the boosted MED, and the first effect and the liquid–liquid heat exchanger of the flash-boosted MED processes for 75°C heat source inlet temperature scenario.

heat transfer includes both sensible (increasing the feed inlet temperature from 38°C to the relevant boiling point temperature) and latent heat transfers (that relates to the evaporation at the vapor temperature that is slightly lower than the boiling point temperature due to the boiling point elevation (BPE)). The booster temperature-energy profile is similar to the first effect profile due to their same nature of operation. However, in the brine heater of the FB-MED process, the heat source energy is transferred to the recirculated brine stream with an approach of 3°C for the heater as shown in Fig. 8.7, which is released gradually within the flashing chambers and injected to the relevant effects of the primary MED section to bring about a 43% higher production yield as compared to the conventional MED process. Overall, the FB-MED process has a higher potential to exploit the heat source energy as compared to other processes (namely, conventional MED and B-MED). As it shows, for 75°C heat source inlet temperatures, the FB-MED can exploit 21% and 45% more energy from the heat source as compared to the B-MED and the conventional MED, respectively.

The trend of the energy transferred over all sections of the processes (MED, B-MED, and FB-MED) is shown in Fig. 8.8 for a 75°C heat source inlet temperature scenario. Because of the higher temperature drop over the heat source stream at the first effect of the conventional MED (Fig. 8.7), the amount of energy released in the first effect of the conventional MED is around 192% more than that at the first effect of the FB-MED

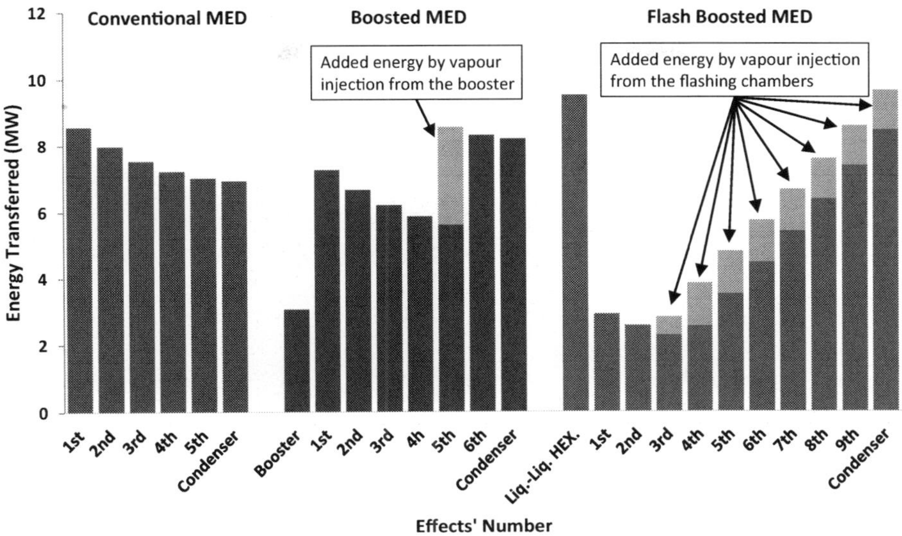

Figure 8.8
Energy released of each effect of all three systems for 75°C heat source inlet temperature scenario.

system at 75°C heat source inlet temperature. In contrast, it is around 18% more than that at the first effect of the B-MED system. The amount of released energy decreases from effect to effect for the conventional MED. The same happens for B-MED, except at the fifth effect, where steam injection occurs. For FB-MED, this decrement continues until the third effect (i.e., before injection occurs) and, starting from the fourth effect, the energy increases due to steam injection and this trend continues to the end of the process.

As shown in Fig. 8.8 the total amount of released energy in the primary MED section of the FB-MED system is more than that of the conventional MED system and as mentioned before, this extra energy is realized through the flashing chambers, which effectively process more energy from the heat source by using the liquid–liquid heat exchanger (first column as shown in Fig. 8.8 for FB-MED). The same happens for B-MED, however the amount of released energy from the booster section is much lower (3.1 times less) than the amount of released energy in the liquid–liquid heat exchanger of the FB-MED process, which results in a lower improvement as compared to the FB-MED system.

Fig. 8.9 shows the *UA* value distribution for all the considered systems for 75°C heat source inlet temperature scenario. For the FB-MED's liquid–liquid heat exchanger, the heat transfer rate is around 9.53 MW (Fig. 8.7); coupled with a constant approach temperature of 3°C for both sides of the heat exchanger (Table 8.1), 3.18 MW/°C *UA* value is required. Since the major part of heat transfer that occurs in both the first effect, and in the condenser of all configurations (conventional MED, B-MED, and FB-MED) pertains to sensible-latent heat transfer instead of latent–latent heat transfer as in other

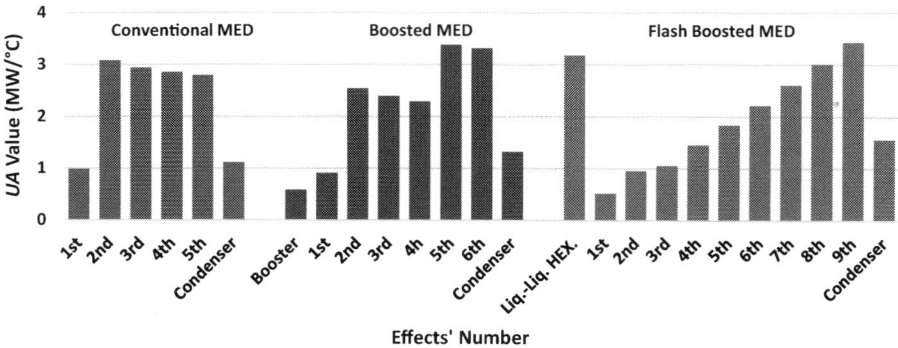

Figure 8.9
UA-value of each effect of all three systems for 75°C heat source inlet temperature scenario.

effects, this results in higher *UA* values for those effects compared with those of the first effect, the booster and the condenser.

Referring to Chapter 5, the pumping power and normalized pumping power consumptions of all processes are shown in Figs. 8.10 and 8.11, respectively. It demonstrates that the FB-MED process has the highest auxiliary power consumption, which is mainly related to the required pumping power for the plate-type liquid—liquid heat exchanger (the brine recirculation and the heat source medium pumps). As mentioned before, the number of required plate heat exchanger in each case is calculated based on the required *NTU*; then these plate heat exchangers are connected together in series to maintain the required temperature drop. The maximum *NTU* per plate heat exchanger is 5.0 [5]. As shown in

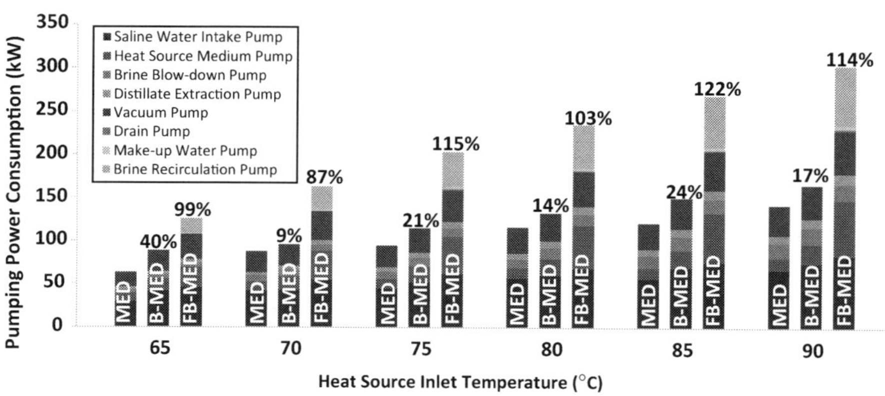

Figure 8.10
Pumping power (kW) versus heat source inlet temperature (°C) of the optimized conventional multi-effect distillation (MED), boosted MED, and the flash-boosted MED processes. (The values are indicative of the percentage ratio as compared to the conventional MED alternative for each specific heat source temperature).

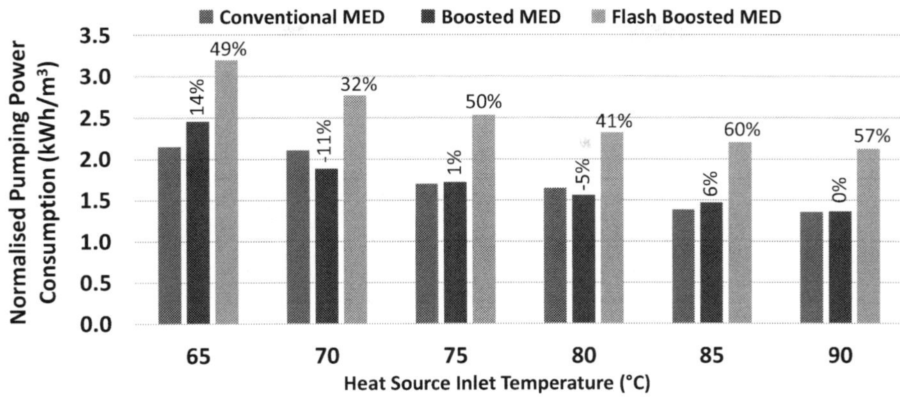

Figure 8.11

Normalized (specific) pumping power (kWh/m^3) versus heat source inlet temperature (°C) of the optimized conventional multi-effect distillation (MED), boosted MED, and the flash-boosted MED processes. (The values are indicative of the percentage ratio as compared to the conventional MED alternative for each specific heat source temperature.)

Fig. 8.10, the pumping power consumption of the FB-MED is roughly double that of the conventional MED. However, due to the higher yield of FB-MED as compared to the conventional MED, the normalized pumping power increment is reduced to a range between 30% and 60% (Fig. 8.11). In contrast, the pumping power increment of B-MED is in a reasonable range between 10% and 40%, which brings about a substantial reduction in the normalized pumping power and is in the range of −11% to 14% (Fig. 8.11).

The impact of pumping power consumption appears in the operating costs, which will be discussed in the economic analysis.

8.3 Thermoeconomic Analysis

Stemming from the work of Wang et al. [3] and Rahimi et al. [2,9], into B-MED and FB-MED we will optimize the conventional MED, B-MED, and FB-MED design cases for the production rate using the methods as described in Chapter 7. We will assess whether the FB-MED process could engender a greater economic return than plants utilizing the conventional MED or the B-MED process. Table 8.4 specifies all the significant assumptions used in conducting the present analyses.

A cash flow analysis was performed upon the annual cash flows of each MED, B-MED, and FB-MED plant design, in a manner described in the prior section. Each unique plant design can be characterized by the temperature of heat source fluid, the number of primary MED effects utilized, and the number of flashing chambers utilized (which is applicable for FB-MED only).

Table 8.4: Assumptions of our economic analyses [9].

Plant availability, *f*	95%
Plant useful life, *yr* (years)	30
Interest rate, *i* (%)	8
Capital recovery factor, *CRF*	0.089
Plant residual value at the end of useful lifetime (%)	0
Tax rate (% of taxable income)	30
Percentage of capital cost covered by loan (%)	50
Additional charge on loan as bank fee (% of capital cost)	2
Water market price, *WMP* (AU$/m^3)	2.72 [10]
Annual escalation factor on incomes and OPEX (%)	3

8.3.1 Capital Costs

The capital costs of FB-MED and B-MED plants are greater than the MED alternative. However, due to the higher production yield of these processes as compared to the conventional MED, their normalized capital costs are lower than that of the conventional MED (Fig. 8.12).

As shown in Fig. 8.12, the normalized capital cost of both B-MED and FB-MED processes are generally 3% less for FB-MED and 1% less for B-MED, as compared to the MED alternative for all scenarios.

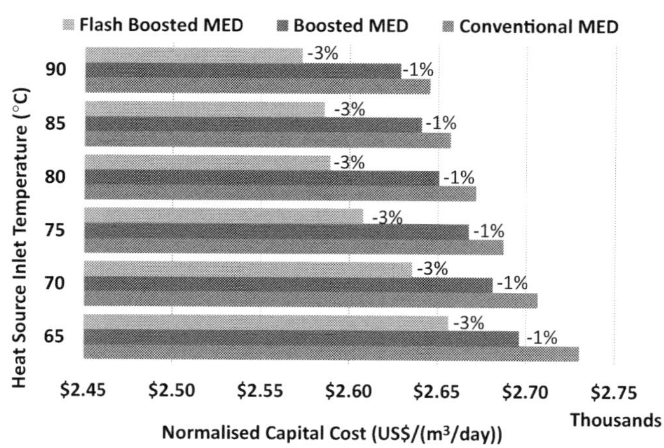

Figure 8.12

A comparison of the normalized capital costs of optimized multi-effect distillation (MED), boosted MED, and flash-boosted MED plants for each considered heat source temperature. The values are indicative of the percentage decrease in the specific capital cost of the boosted MED or flash-boosted MED, as compared to the conventional MED alternative for each specific heat source temperature.

8.3.2 Operating Costs

The operating costs include the costs of electricity, labor, chemical additives, maintenance, spare parts, and insurance (see Chapter 7). Our results show that the annual operating cost increases with heat source fluid temperature and is greater for both B-MED and FB-MED than the operating cost of the MED alternative. The general increase in operating cost can be attributed to several factors such as auxiliary pumping power, and are associated with the increased production rates of the FB-MED and B-MED.

As Fig. 8.13 shows, the electrical expenses constitute the greatest increase in the operating expenditure for the FB-MED system. The FB-MED system has several more pumps to handle flows that enable the increased thermal efficiency of the system; this can be seen by comparing Figs. 8.1–8.3. The make-up water, brine recirculation, and drain pumps can be seen in Fig. 8.3, the operation of which are contributing extra duty requirement to the FB-MED system. Furthermore the pumps that exist in both MED and FB-MED plants have a greater duty in the FB-MED system. The heat source pump has a greater duty requirement because of the greater pressure difference due to the extra heat exchanger (Fig. 8.3). Due to the increased production of the plant, the brine, freshwater, NCG (noncondensable gas) extraction, and saline water pumps also have increased duties, and therefore greater electrical consumption.

Despite the overall increase in operating costs, the specific operating costs for B-MED and FB-MED processes are comparable to that of the MED process, as can be seen in

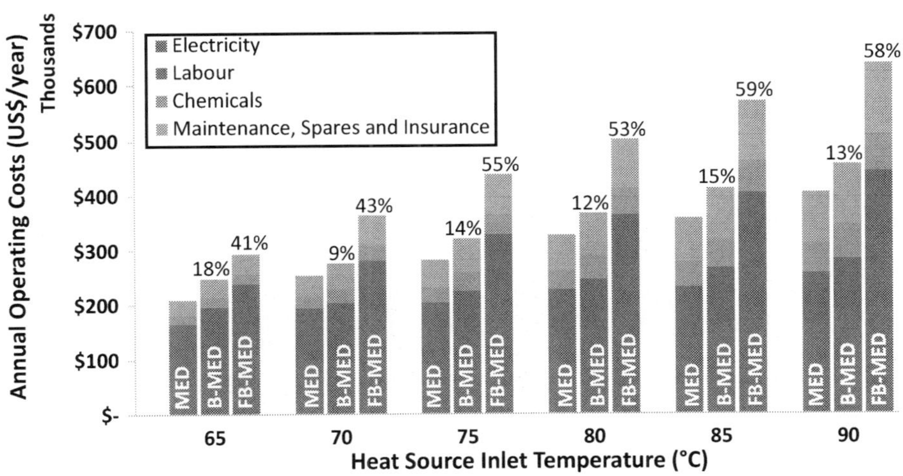

Figure 8.13

Breakdown of the annual operating costs versus heat source fluid inlet temperature for the optimized conventional multi-effect distillation (MED), boosted MED, and flash-boosted MED processes. The percentage increase in total annual operating cost over the MED alternative is noted for the boosted MED and flash-boosted MED for each heat source inlet temperature.

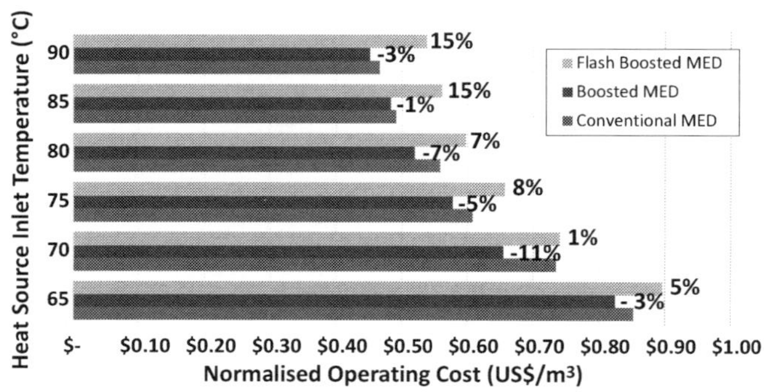

Figure 8.14

Normalized annual operating costs of optimized multi-effect distillation (MED), boosted MED, and flash-boosted MED plants for each considered heat source temperature. The values beside each bar are indicative of the percentage difference in the specific operating cost of the boosted MED or flash-boosted MED, from the conventional MED alternative for each heat source inlet temperature.

Fig. 8.14. This is due to the higher production rates of B-MED and FB-MED as compared to the conventional MED. As shown in Fig. 8.14, B-MED has a lower operating cost in all scenarios; in contrast the operating cost of the FB-MED process is higher than the conventional MED and the extent ranges from 1% to 15%.

8.3.3 Unit Product Cost

Fig. 8.15 details the trend of unit product cost (*UPC*) for B-MED and FB-MED compared to the conventional MED process. The method to find the *UPC* over the plant's lifetime is based on the net present value (*NPV*) analysis. In this method the *UPC* is equal to the water

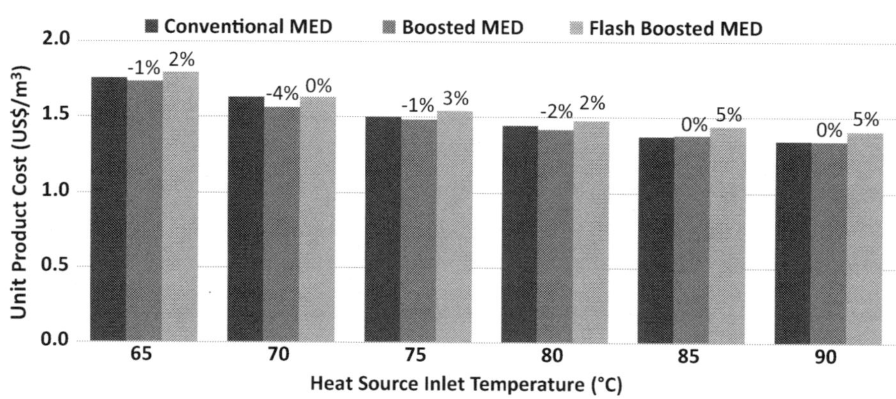

Figure 8.15

UPC of the optimized multi-effect distillation (MED), boosted MED, and flash-boosted MED plants for each considered heat source temperature. The values above each column are indicative of the percentage difference in *UPC* of the boosted MED or flash-boosted MED, to the conventional MED alternative for each specific heat source temperature.

market price (see Section 7.4.4), which makes for a zero *NPV* over the plant's lifetime considering all costs including operating, depreciation, loan repayment, taxes, and so forth.

As shown in Fig. 8.15, the *UPC* of B-MED and FB-MED processes is comparable and is approximately in the range of ±5% of the conventional MED process.

8.3.4 Cash Flow Analysis

As explained in Chapter 7, the *NPV* of the alternatives is calculated for a range of assumed interest rates (discount rate) for all scenarios. Fig. 8.16 shows this graph for 80°C heat source inlet temperature (°C). As it shows, for any interest rate less than 17.4% the FB-MED is the superior option, the B-MED is the best option for the interest rates between 17.4% and 19.7%, and for any interest rate more than 19.7%, none of them is profitable.

Table 8.5 lists the best options for the various scenarios. According to the assumed conditions (see Table 8.1), since the interest rate is 8%, the FB-MED is the best choice in all the scenarios.

As it shows in Table 8.5, for an 85°C inlet heat source temperature, the Δ*IRR* between FB-MED and B-MED is 17.5% which means that for any interest rate less than that, the FB-MED has the highest *NPV* and should be selected as the best choice. By increasing the interest rate it shows that at 20.0% the *NPV* of MED is greater than that of the B-MED, therefore B-MED should be selected between 17.5% and 20.0%.

Fig. 8.17 shows the impact of each flashing chamber in the FB-MED process. In this figure, the *NPV* of zero flashing chamber is relevant to the optimized conventional MED for that scenario. As it shows, by adding the flashing chambers the *NPV* in each scenario

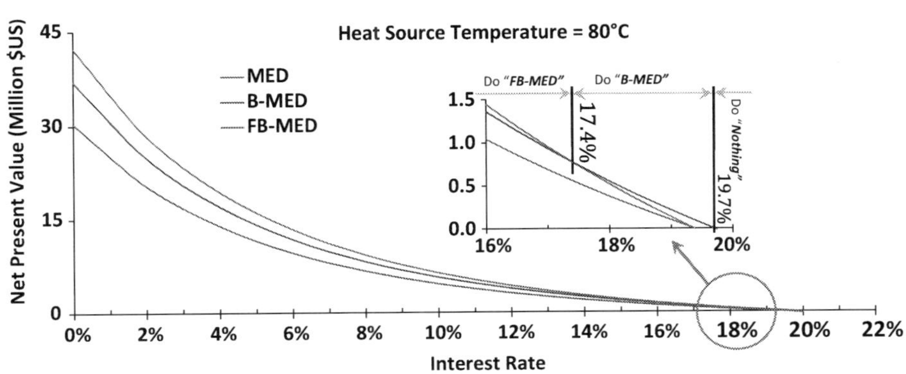

Figure 8.16

Net Present Value (*NPV*) versus interest rate plots for the optimized conventional multi-effect distillation (MED), boosted MED, and flash-boosted MED plant designs for 80°C heat source inlet temperature.

Table 8.5: The best option over the interest rate range for all scenarios.

Inlet heat source temperature (°C)	Best choice over the interest rate range
65	0% FB-MED 9.1% B-MED 16.4% Nothing
70	0% FB-MED 13.7% B-MED 18.2% Nothing
75	0% FB-MED 16.5% B-MED 19% Nothing
80	0% FB-MED 17.4% B-MED 19.7% Nothing
85	0% FB-MED 17.5% B-MED 20.0% MED 20.2% Nothing
90	0% FB-MED 17.1% B-MED 20.5% Nothing

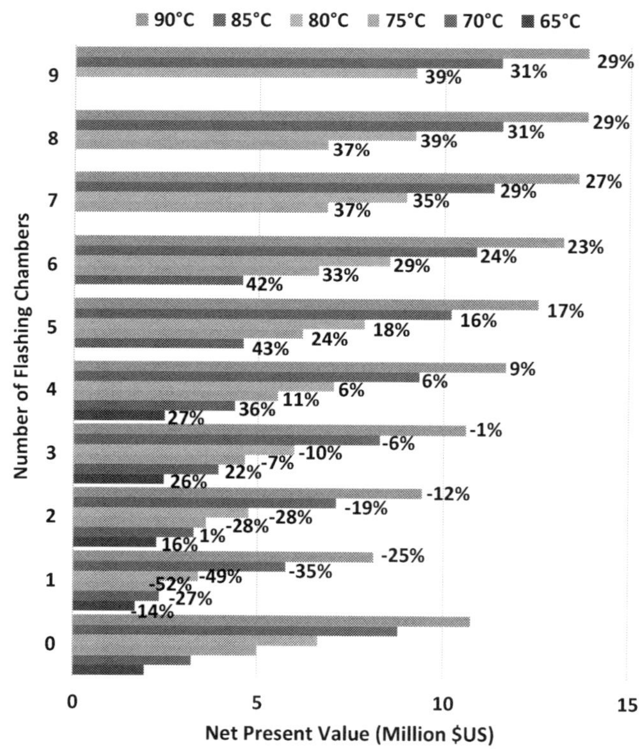

Figure 8.17

The *NPV* versus number of installed flashing chambers for all considered temperatures of the heat source stream. Those plant designs that utilize zero flashing chambers are conventional multi-effect distillation (MED) plants, and all remaining plant designs are flash-boosted MED. The percentage values beside the bars indicate the percentage increase in *NPV* for that flash-boosted MED design compared with the conventional MED alternative (0 number of flashing chamber) for that particular heat source temperature.

Table 8.6: The optimized number of effect and flashing chamber for flash-boosted multi-effect distillation in each scenario.

Inlet Heat Source Temperature (°C)	Number of Effects	Number of Flashing Chamber
65	5	4
70	7	5
75	9	7
80	10	9
85	10	8
90	10	9

is increasing, except for the last two possible chambers (8 and 9), for which the *NPV*s are roughly equal. Table 8.6 shows the optimized number of flashing chamber based on the result of Fig. 8.17.

Fig. 8.18 compares the optimized configurations of all MED, B-MED, and FB-MED plant designs and indicates the incremental improvement in *NPV* over the MED, according to the available market price for water, that is 2.50 US\$/m^3 (equivalent of 2.72 AU\$/m^3 [10]). As it shows, the FB-MED process has a considerably greater *NPV*, being up to 43% greater than that of the MED alternative. Similarly, the *NPV* of B-MED installations can be up to 32% greater than that of conventional MED. We also note that *UPC*, in general,

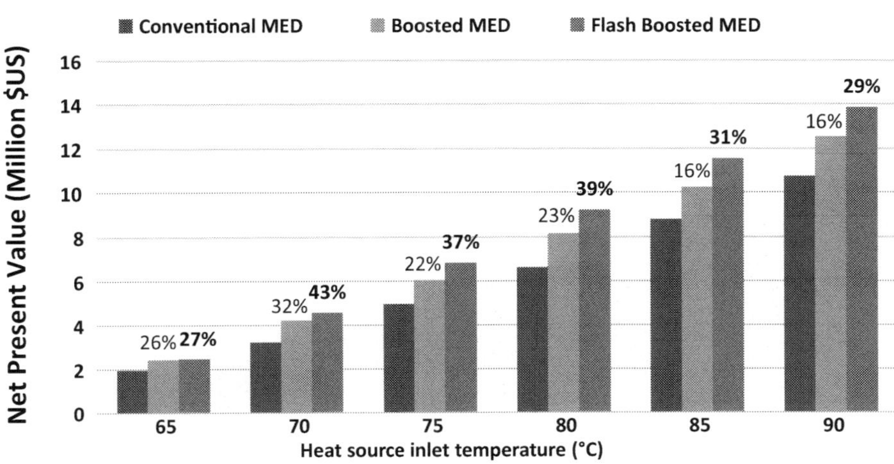

Figure 8.18

Net present value of the optimized configurations of all multi-effect distillation (MED), boosted MED, and flash-boosted MED plants for each considered heat source temperature. The values above each column are indicative of the percentage increase in *NPV* of the optimized boosted MED and flash-boosted MED over the optimized conventional MED alternative for a specific heat source temperature.

cannot clarify the profitability of two systems [9]. Referring to both Figs. 8.15 and 8.18, at 65°C heat source inlet temperature, the FB-MED has the highest *NPV* and the highest *UPC*.

8.4 Conclusion

Three configurations—the conventional MED, B-MED, and FB-MED desalination processes—have been simulated. The FB-MED displays the highest production rates and performance ratios over an extensive range of low grade sensible heat source temperatures. A generalized method for estimating the costs of conventional MED, B-MED, and FB-MED plants has been developed and numerous configurations have been analyzed accordingly. Measures of profitability have been obtained for each design so that the economic viability of FB-MED designs can be assessed in terms of the number of installed flashing chambers. The results indicate that the increased production capacity and thermal efficiency of the B-MED and FB-MED processes outperform the increased capital and operational requirements inherent to these processes. Results of the analysis indicate that the optimized B-MED and FB-MED systems can have an *NPV* that is up to 32% and 43%, respectively, greater than the optimized conventional MED, depending on the temperature of the utilized heat source. We have also shown that the *NPV* is a better measure of profitability than *UPC*, as *NPV* is able to differentiate multiple propositions.

References

[1] A. Christ, K. Regenauer-Lieb, H.T. Chua, Boosted Multi-Effect Distillation for sensible low-grade heat sources: a comparison with feed pre-heating Multi-Effect Distillation, Desalination 366 (2015) 32–46.

[2] B. Rahimi, A. Christ, K. Regenauer-Lieb, H.T. Chua, A novel process for low grade heat driven desalination, Desalination 351 (October 2014) 202–212.

[3] X. Wang, A. Christ, K. Regenauer-Lieb, K. Hooman, H.T. Chua, Low grade heat driven multi-effect distillation technology, Int. J. Heat Mass Transfer 54 (25–26) (December 2011) 5497–5503.

[4] A. Christ, K. Regenauer-Lieb, H.T. Chua, Application of the Boosted MED process for low-grade heat sources — a pilot plant, Desalination 366 (2015) 47–58.

[5] C. Haslego, G. Polley, Compact heat exchangers – Part 1: Designing plate-and-frame heat exchangers, Chem. Eng. Prog. 98 (9) (2002) 32–37.

[6] A. Christ, K. Regenauer-Lieb, H.T. Chua, Thermodynamic optimisation of multi-effect-distillation driven by sensible heat sources, Desalination 336 (2014) 160–167.

[7] B. Rahimi, H.T. Chua, A. Christ, System and Method for Desalination, World Intellectual Property Organization, 2015. WO 2015/154142 A1.

[8] H.T. Chua, K. Regenauer-Lieb, X. Wang, A Desalination Plant, World Intellectual Property Organization, 2012. WO 2012/003525 A1.

[9] B. Rahimi, J. May, A. Christ, K. Regenauer-Lieb, H.T. Chua, Thermo-economic analysis of two novel low grade sensible heat driven desalination processes, Desalination 365 (2015) 316–328.

[10] Australian Bureau of Statistics (ABS), Western Australia. State Fact Sheet 2013, cited 8 May 2014, 2014 [Online]. Available: http://www.abs.gov.au/AUSSTATS/.

CHAPTER 9

Application of Novel Low Grade Heat-Driven Distillation in Alumina Refineries

9.1 Introduction

Alumina itself is the precursor to aluminum, which is 8% of the earth by mass [1] and is typically found in the form of bauxite. Aluminum is the most abundant metal and the second most produced metal in the world, after steel [2,3]. Bauxite is the essential raw material to produce alumina. It contains approximately 30% alumina in the form of aluminum hydroxide and Australia is the world leading producer of it. In 2012, Australia produced 79.4 Mt of bauxite from five bauxite mines, which was 65% more than China, the second largest bauxite producer [4].

Alumina is produced via the Bayer process, which was invented and patented in 1887 by Karl Joseph Bayer [1], and it has since changed very little over a century [5]. China with 20.3 Mt of alumina in 2012 remains the world's largest alumina producer [4]. In the Bayer process the use of hot caustic soda is essential because it dissolves alumina, allowing the dissolved solution to be separated from undissolved waste for further processing. In this process, crushed bauxite is subjected to high-temperature digestion in a concentrated caustic solution. The resulting liquor (pregnant liquor), is then clarified and filtered to remove the mud and other insoluble residues. The clarified pregnant liquor is then cooled and seeded with solid aluminum hydroxide to precipitate the alumina. The resulting spent liquor is concentrated by an evaporation process and recycled for further bauxite digestion [5–7], which is an important part of the Bayer process in alumina refinery plants, both to make water for the residue washing process, and to reconcentrate the spent liquor for the bauxite digestion process.

The evaporation process consumes a significant amount of thermal energy. The evaporation section is also responsible for controlling the water balance in the whole plant. Its role can therefore be summarized as balancing water and discharging impurities. Its main heat source is the available low pressure live steam and as compared to the other sections of the plant, it is the most energy-intensive process, accounting for approximately 25–30% of energy consumption per ton of alumina [8]. Alumina refineries are large users

Low Grade Heat Driven Multi-Effect Distillation and Desalination. http://dx.doi.org/10.1016/B978-0-12-805124-5.00009-7

of gas, coal, and other relevant fuels for their energy requirements. For example in Australia in 2012, alumina refineries consumed 221 PJ of energy [4]. Any improvement to this section that can decrease steam consumption, which in turn saves fuel consumption in the steam plant, will significantly reduce greenhouse gas emission and production cost. In 2012, 0.7 tonnes of CO_2-e was produced per tonne of alumina in Australia [4].

9.2 Evaporation Process

The design of an evaporator is purely based on its suitability to handle process liquor, which is viscous and prone to scaling. Therefore, both tube falling film and flashing vessel techniques (evaporators) can be used. Recently the plate falling film evaporator is also used by the Aluminum Company of Shanxi and Guizhou Alumina [9,10] to improve the evaporation efficiency in evaporation sections with low concentration feed streams. It is worthwhile to mention that a state-of-the-art pilot plant (see Chapter 3) onsisting of two serially connected Alfa Laval's single-effect rising film plate evaporator/condenser modules [11] is also being tested in collaboration with the Worsley Alumina Refinery Plant to study and address the potential scaling issues when it is coupled to different process streams (Fig. 9.1).

Both flashing and evaporation techniques are shown in Figs. 9.2 and 9.3. As shown in Fig. 9.2, a combination of heat exchangers and flash vessels known as multi-stage flash (MSF) can be used. In this system, the feed liquor is first preheated by going through a series of heat exchangers (preheaters) and is then introduced to a battery of flash vessels. The flashed vapors from the vessels are condensed in the preheaters, and the concentrated

Figure 9.1
The two serially connected Alfa Laval's single-effect rising film plate evaporators modules (JWP-16-C Series) [11].

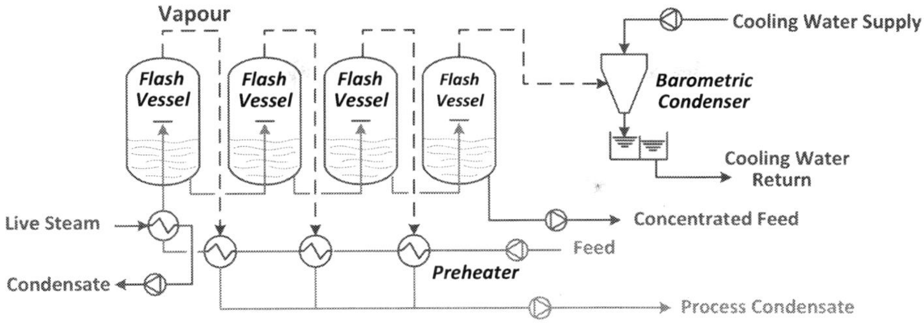

Figure 9.2
Schematic design of a multi-stage flash evaporation unit in alumina refineries.

feed from the last outlet is the final product of the evaporation unit. The first heat exchanger (known as the brine heater in MSF desalination plants) in this system uses the available live steam as the heat source.

Instead of flash vessels, the falling film technique with a higher thermal performance [12] can be used. However, it is more prone to scaling, especially at the higher temperature as compared to the MSF process, and requires periodic in-line cleaning. In this technique (Fig. 9.3) the feed liquor distributes onto the evaporator from the top. Gravity pulls the film down along the inner walls of the evaporator tubes, so that the fast moving and thinning film gives rise to a high heat transfer coefficient [13]. At the bottom of the heat exchanger the concentrated feed stream is collected. The outgoing vapor from the first effect can be the heat source for the second effect, and this trend continues through to the last condenser, which makes up a multi-effect evaporation (MEE) process. Independent of the evaporator types, an MEE process is the preferred choice to reduce energy consumption.

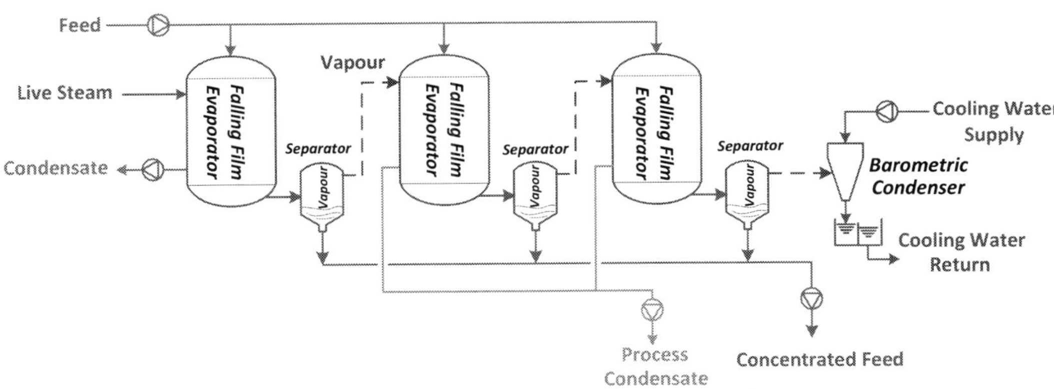

Figure 9.3
Schematic design of a multi-effect evaporation unit in alumina refineries.

The aim of this chapter is to introduce two novelties (see Sections 9.3 and 9.4). The first is a novel process that can be coupled with sensible waste heat sources in an alumina refinery plant to reduce the steam consumption of the main evaporation unit [14]. This reduces fuel consumption, greenhouse gas emission, and production cost (Section 9.3).

The second novelty (Section 9.4) introduces a process that can ultimately be coupled with the produced vapor in the last flash vessel of the conventional MSF process that is injected into the barometric condenser and is mixed with the cooling water supply stream [15]. By using these available waste heat streams in the evaporation units of an alumina refinery plant, the live steam consumption (which is the main heat source of the evaporation process) can be reduced dramatically.

It should be clarified that the novelties herein lie with the process design instead of the process units themselves; the latter are simply standard technologies. The fact that standard process units are used augurs well for scale-up adoption in refineries. For illustration purposes, we elect to focus on some available waste heat streams that come out from one or more evaporation units of the Worsley Alumina Refinery Plant. For example, in the first novelty (Section 9.3), this stream is the outlet process condensate stream of the main evaporation unit (Fig. 9.2). This novel technique can potentially harvest all available sensible waste streams to run an auxiliary evaporation unit so as to reduce the load of the main evaporation unit, which in turn saves the steam consumption and therefore coal consumption of the Worsley Alumina Refinery Plant. In the second novelty, the low grade produced flashed vapor that exits from the last flashing chamber (Fig. 9.2) is used to run a developed thermal vapor compression multi-effect evaporation (TVC-MEE) process (Section 9.4).

9.3 A Novel Flash-Boosted Multi-Effect Evaporation for Alumina Refineries

A flash-boosted MEE (FB-MEE) process is coupled with low grade waste sensible heat streams to boost freshwater production as compared to the conventional MEE process. This is the equivalence of the flash-boosted multi-effect distillation (FB-MED) process in mineral refinery applications.

9.3.1 Process Description

Fig. 9.4 shows the process when it is coupled with the outlet process condensate of the evaporation unit (Fig. 9.2). The available temperature is around 85°C. The simulation has also been done for 75°C to cover the minimum available temperature as well.

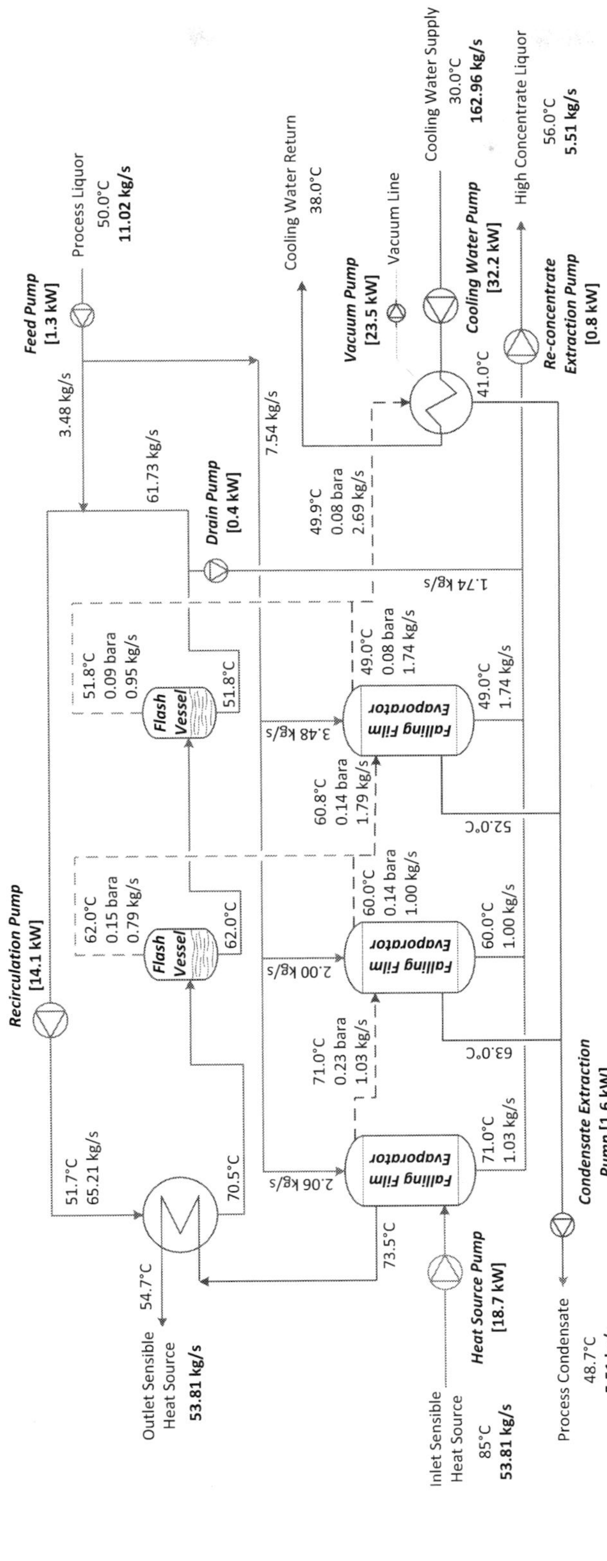

Figure 9.4
Schematic design of a flash-boosted multi-effect evaporation unit (parallel feed; three falling film effects and two flash vessels) coupled with an 85°C low grade heat source.

As shown in Fig. 9.4, the process condensate (inlet sensible heat source) is first used to power the primary MEE section, and then to heat the feed stream that supplies the flash vessels. Each vessel produces flashed vapor, which is directed to an appropriate primary MEE effect according to the relevant pressure differences, thereby supplying more heat and increasing the amount of feed that can be evaporated across the MEE effect. The result of this vapor injection scheme is an increase in freshwater production and reconcentrate flow rate across the primary MEE effects. The number of falling film evaporation effects can be set based on the available temperature difference between the heat source and the cooling water temperature and the relevant boundary conditions [16,17].

Fig. 9.5 shows the schematics of the conventional MEE coupled with an 85°C low grade heat source as the benchmark. The higher temperature drop for the heat source in the case of the FB-MEE process as compared to the conventional MEE process means that more waste heat has been effectively utilized to improve the production rate.

9.3.2 Numerical Analysis and Validation

A simulation model has been developed to quantify the efficiency of the FB-MEE process [16,18]. The mathematical model includes mass, concentration, and energy balances for each effect. Mass and concentration balances are written as $k \in \{1,\ldots,n\}$:

$$\dot{m}_{F,k} = \dot{m}_{V,k} + \dot{m}_{HC,k} \tag{9.1}$$

$$\dot{m}_{F,k} \cdot X_F = \dot{m}_{HC,k} \cdot X_{HC} \tag{9.2}$$

The energy balance in the first effect can be written as

$$\dot{Q}_1 = \dot{m}_{F,1} \cdot h_{f_{F,1}} + \dot{m}_{HS} \cdot h_{f_{HS,in}} = \dot{m}_{HS} \cdot h_{f_{HS,out}} + \dot{m}_{V,1} \cdot h_{g_{(T_{HC,1},P_{sat,1})}} + \dot{m}_{HC,1} \cdot h_{f_{HC,1}} \tag{9.3}$$

$\dot{Q}$ is the total heat transfer (kW), $\dot{m}_F$, $\dot{m}_{HS}$, $\dot{m}_V$, and $\dot{m}_{HC}$ are the feed, heat source, generated vapor, and reconcentrated feed mass flow rates (kg/s), respectively. h_f and h_g are the enthalpies of saturated liquid and steam (kJ/kg), respectively.

Figs. 9.6 and 9.7 illustrate the temperature—energy profile across the first effect through to the last effect and the condenser of both conventional MEE and FB-MEE processes for 85°C heat source inlet temperature. The process conditions considered conform to the assumptions listed in Table 9.1.

In the first effect, the heat source temperature drops across the evaporator, while the feed liquor temperature first increases from the inlet temperature to the relevant boiling temperature (T_{HC}) (preheating zone) according to its concentration at the prevailing pressure (P_{sat}), and then increases further during the evaporation process because of continuing concentration that brings about heightened boiling point elevation (*BPE*)

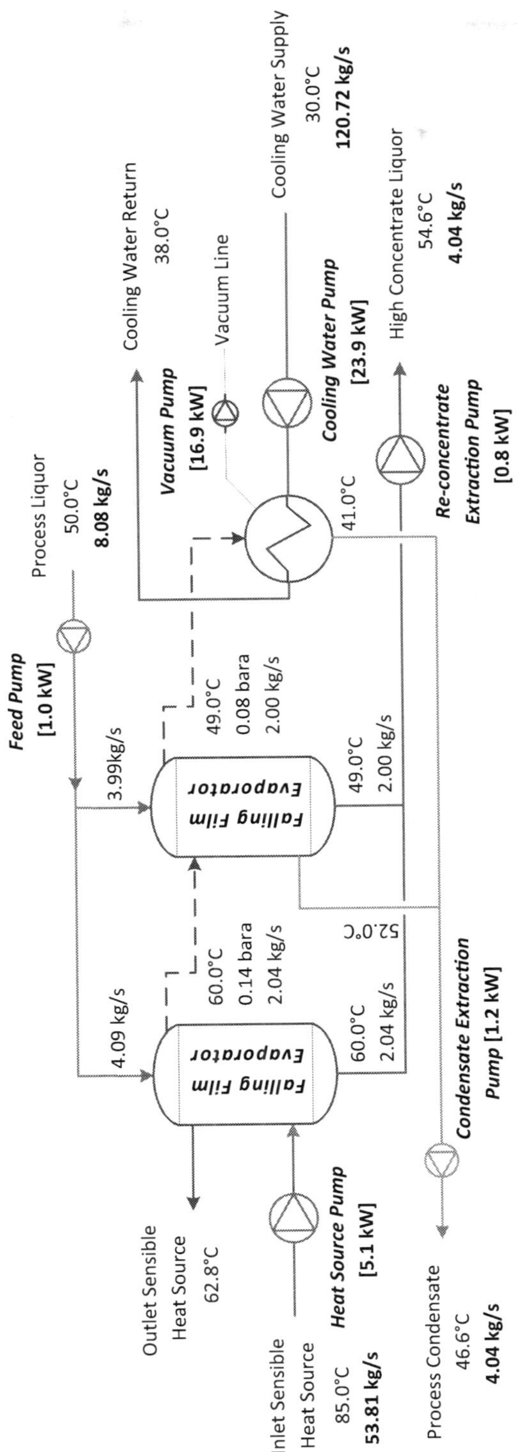

Figure 9.5
Schematic design of a conventional multi-effect evaporation process coupled to an 85°C low grade heat source.

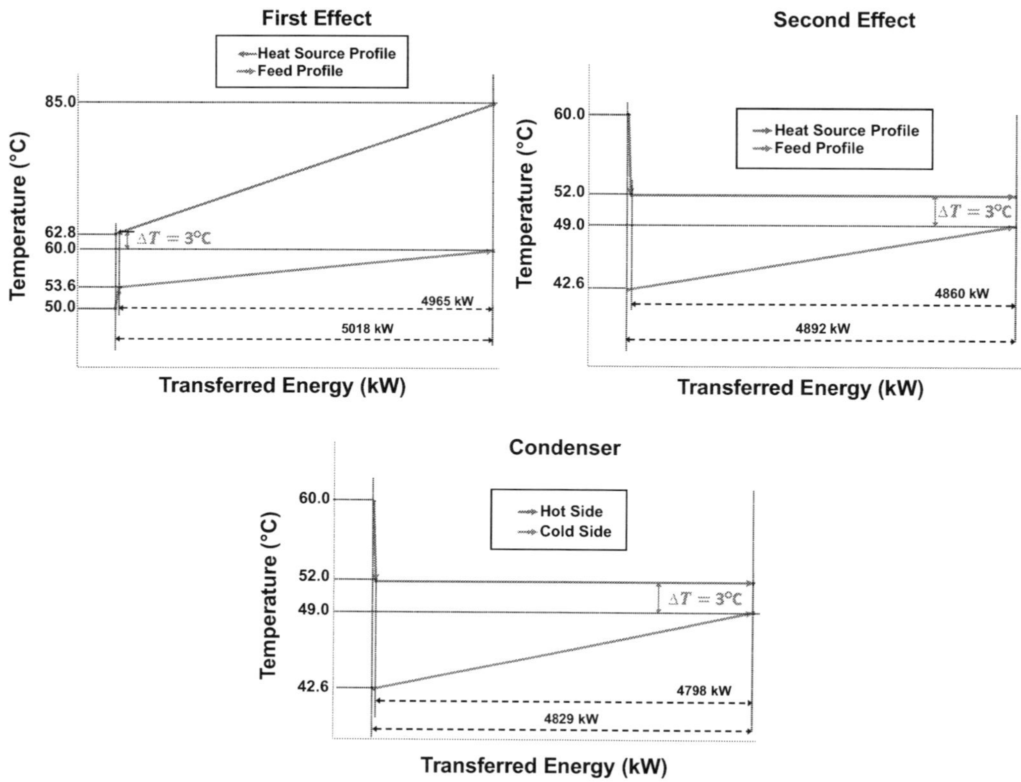

Figure 9.6
Temperature-energy profiles of the conventional multi-effect evaporation process for 85°C scenario.

(evaporation zone). Therefore due to the different types of heat transfer between preheating and evaporation zones, there are two different *UA* values. In this case, the preheating zone covers less than 4% (Figs. 9.6 and 9.7) of the total energy transferred and therefore to a good approximation it can be ignored. Hence the overall *U* (heat transfer coefficient, ($kW/m^2 \cdot K$)) value for the first effect pertains to that of the evaporation zone.

For the second effect, $k \in \{2,\ldots,n\}$, where n is the total number of effects, we have

$$\begin{aligned}\dot{Q}_k &= \dot{m}_{V,k-1} \cdot h_{g_{(T_{HC,k-1},P_{sat,k-1})}} + \dot{m}_{F,k} \cdot h_{f_{F,k}} \\ &= \dot{m}_{V,k-1} \cdot h_{f_{D_{(T_{sat,k-1})}}} + \dot{m}_{V,k} \cdot h_{g_{(T_{HC,k},P_{sat,k})}} + \dot{m}_{HC,k} \cdot h_{f_{HC,k}}\end{aligned} \tag{9.4}$$

h_{f_D} is the enthalpy of distillate at its saturation temperature (kJ/kg). Because the heat source of these effects is the superheated vapor from the previous effect, the superheated vapor desuperheats to the relevant saturated temperature of the previous effect, and then it condenses. In contrast, the trend of feed temperature profile is the same as in the first

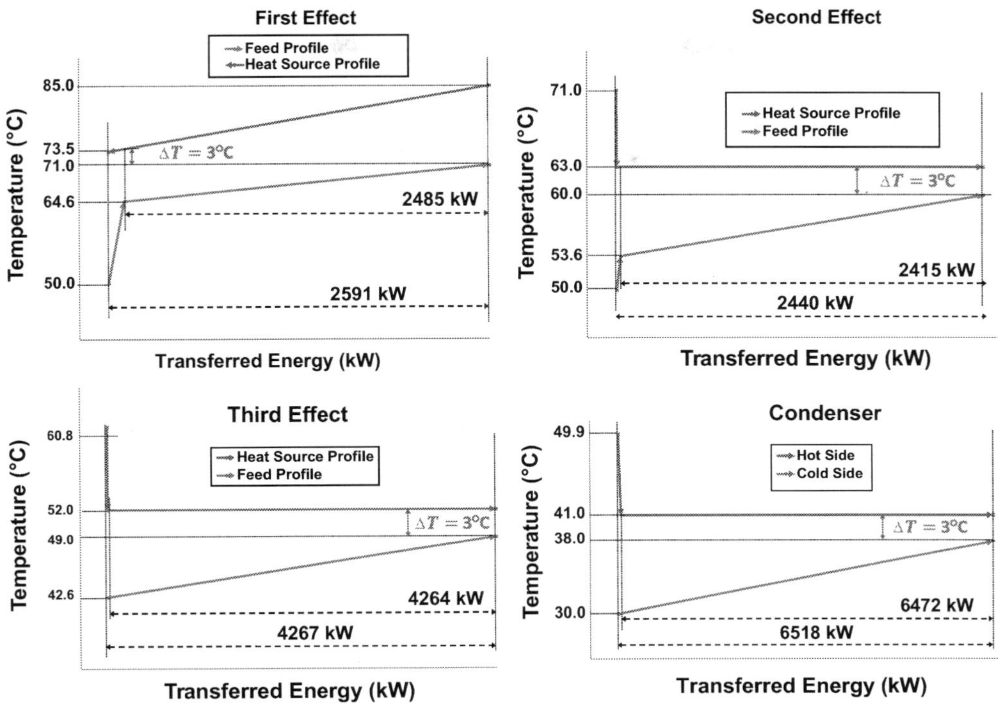

Figure 9.7
Temperature-energy profiles of the flash-boosted multi-effect evaporation process for 85°C scenario.

effect if the saturation pressure of the feed stream is less than the effect's pressure, as with the second effect of the FB-MEE process as shown in Fig. 9.7. Therefore, in this case, the feed temperature increases to the boiling point. The temperature then increases further during the evaporation process due to the changing concentration and the relevant *BPE*. In

Table 9.1: Assumptions of thermodynamic simulations.

Heat source flow rate, $\dot{m}_{HS,1}$ (m^3/hr)	200
Heat source temperature range (°C)	75, 85
Feed-to-vapor ratio (50% evaporation)	2.0
$T_{C,in}$ (°C)/$T_{C,out}$ (°C)	30/38
ΔP_{inj}[a] (Pa)	1000
Liquid-to-liquid heat exchanger approach temperature (°C)	3.0

[a]Relevant pressure difference for steam injection purpose in flash-boosted multi-effect evaporation processes.

this case, to an excellent approximation, the overall U value for these effects is simply that of the evaporation zone.

Sometime the feed saturation pressure is more than the effect's pressure, as in the second effect of the conventional MEE and the third effect of the FB-MEE process. In this case, before the feed is being distributed into the tubes, a negligible amount of feed (less than 1%, which corresponds to about 7°C of sensible heat in our case) is flashed as the pressure and temperature decrease. For these effects the inlet feed temperature is considered to equate to the relevant feed boiling temperature (which is related to its concentration) that corresponds to the pressure in the second effect and, in Figs. 9.6 and 9.7, is equal to 42.6°C. After temperature reduction, once it distributes into the tubes, evaporation occurs. As shown in Figs. 9.6 and 9.7, for these effects the amount of energy required for desuperheating is negligible.

For the second and third effects of both processes, the area needed for desuperheating has a significant effect on the capital cost of the evaporator. This will be discussed in detail in Section 9.3.5. In this case because the degree of superheating is less than 10°C, desuperheaters cannot practically be used, and additional heat exchanger area has to be factored in Ref. [19].

In the FB-MEE process the impact of the injected vapor from the flash vessels that mixes with the superheated vapor from the relevant upstream MEE effect has to be considered in arriving at the mixed superheated vapor temperature.

For the condenser,

$$\dot{Q}_{cond} = \dot{m}_{V,n} \cdot h_{g_{(T_{HC,n}, P_{sat,n})}} + \dot{m}_{cond} \cdot h_{f_{cond,in}} = \dot{m}_{V,n} \cdot h_{f_{D_{(T_{sat,n})}}} + \dot{m}_{cond} \cdot h_{f_{cond,out}} \quad (9.5)$$

$\dot{m}_{cond}$ is the cooling water mass flow rate (kg/s). As again Figs. 9.6 and 9.7 show the cooling water temperature increases across the condenser but the trend of the inlet superheated vapor temperature profile is the same as in the other effects, so that the desuperheating zone is negligible once again. The overall heat transfer coefficient is practically that of the latent-to-sensible heat transfer zone. Desuperheating requires extra heat exchanger area.

In all the previous equations,

$$T_{sat} = T_{HC} - BPE \quad (9.6)$$

BPE is the Boiling Point Elevation at the outlet of the reconcentrated process liquor in each effect, and is based on onsite specifications. T_{sat} and T_{HC} are the saturation temperature and the boiling temperature of the reconcentrate (°C). BPE generally depends on the temperature and concentration of the reconcentrated solution. In this context, on account of the large internal vapor volume within the falling film tube evaporator and the fact that the bulk of the evaporation happens at the latter part of the evaporator where it enjoys a more substantial temperature difference between the sensible heat source and the feed, we consider a pragmatic situation where the internal tube temperature throughout the entire tube, which essentially represents that of the evaporated vapor, corresponds to the outlet reconcentrate temperature. Hence, in this simulation as a conservative and pragmatic assumption, the temperature difference between the outlet reconcentrate boiling temperature and the heat source temperature at the pinch point for the first effect in both processes is assumed to be 3°C. For the other effects this 3°C temperature difference is assumed between the outlet reconcentrate boiling temperature and the heat source condensation temperature. An 8°C *BPE* has been considered for the reconcentrate, which is equivalent to the *BPE* of 20 wt% NaOH [20–24]. The *BPE* also dictates the degree of superheating, which in this case is 8°C and remains less than 10°C, which is the threshold for deploying desuperheaters [19].

As shown in Table 9.1, for the purpose of simulation, the heat source medium is the process condensate from the evaporation plant at 75°C or 85°C and its flow rate is 200 m^3/h. The inlet and outlet coolant temperatures at the condenser are 30°C and 38°C, respectively, and it is supplied from a recovery lake close to the site. For the liquid-to-liquid heat exchanger in the FB-MEE process the minimum temperature of approach is taken to be 3°C. The inlet feed is 10 wt% process liquor (1.6°C *BPE* [20–24]; see Fig. B.1 in Appendix B), which is reconcentrated to 20 wt% (8.0°C *BPE* [20–24]; see Fig. B.1 in Appendix B). The feed-to-vapor ratio ($R = \dot{m}_F/\dot{m}_V$) for each MEE effect is taken to be 2.0.

For the flash vessel section (that is applicable only to the FB-MEE process), the energy balance equation is expressed as Eq. (9.7) where $i \in \{1,\ldots,j\}$ is the flash vessel number and j is the total number of flash vessels.

$$\dot{m}_{F,i,in} \cdot h_{f_{F,i,in}} = \dot{m}_{V,i} \cdot h_{g_{(T_{F,out,i},P_{sat,i})}} + \left(\dot{m}_{F,i,in} - \dot{m}_{V,i}\right) \cdot h_{f_{F,i,out}} \tag{9.7}$$

$\dot{m}_{F,i,in}$ and $P_{sat,i}$ are the inlet feed mass flow rate (kg/s) and the saturated pressure (kPa) of the ith flash vessel and

$$T_{sat} = T_{F,out} - (BPE + NEA) \tag{9.8}$$

$T_{F,out}$ is the temperature of the outlet feed of the flash vessel. Nonequilibrium allowance (*NEA*) for flash vessels is a function of flash temperature range, saturation temperature, mass flow rate of outgoing concentrate per unit of vessel width, its level inside the flash vessel, and the design of the flash vessel such as vessel length, width, and the orifice type of the transfer device from vessel to vessel. For this application, a fixed 0.5°C *NEA* is considered. Therefore in each flash vessel a fixed 8.5°C (*BPE* + *NEA*) is present [18]. The injection from each flash vessel to the relevant MEE effect is being considered with 1 kPa saturation pressure difference between the vessels (Table 9.1) [18].

The total amount of freshwater generated is the sum total of the condensate from each of the MEE effects, Eq. (9.9).

$$\dot{m}_{D,total} = \sum_{k=1}^{n} \dot{m}_{V,k} + \sum_{i=1}^{j} \dot{m}_{V,i} \tag{9.9}$$

The generalized reduced gradient (GRG) method, which imposes boundary conditions from the thermodynamic laws and operational, technical, and economic constraints, is used to solve the previous equation set [18]. This method has been validated before in Refs. [16,17,25,26].

9.3.3 Production Rate and Waste Heat Performance Ratio

Fig. 9.8 benchmarks the present FB-MEE scheme with the standard conventional MEE process over a range of heat source inlet temperatures. At 85°C heat source inlet temperature, both the freshwater yield and reconcentrated process liquor yield are increased by 36% with the FB-MEE process.

A waste−heat performance ratio (see Chapter 6) (PR_{WH}), Eq. (9.10), which measures the efficacy of waste heat usage is also used to benchmark the performances. The specific enthalpy of evaporation of 2336 kJ/kg is a typical benchmark to translate the waste heat utilized to an equivalent amount of steam used [16]. The denominator represents the maximum exploitable energy of the heat source relative to the lowest available temperature, which in this case is the condenser inlet temperature

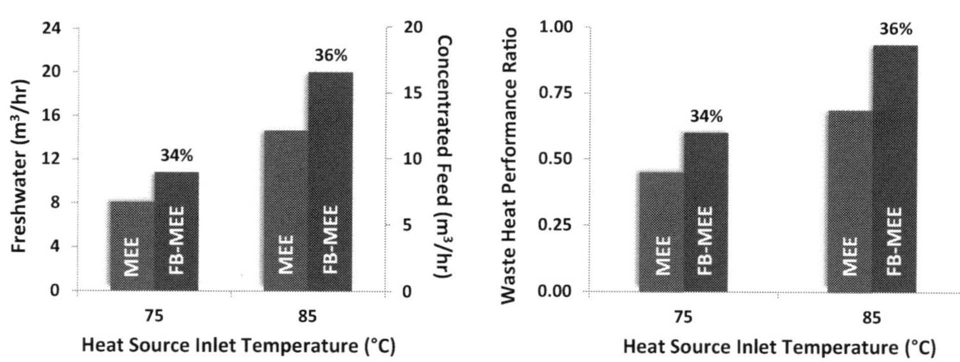

Figure 9.8
Freshwater and concentrated process liquor production rates (left) and waste heat performance ratio for both flash-boosted multi-effect evaporation and conventional multi-effect evaporation systems (right) over the two heat source temperatures.

$$PR_{WH} = \frac{2336 \cdot \dot{m}_{D,total}}{\dot{m}_{HS} \cdot \left(h_{f,HS,in} - h_{f,cond,in}\right)} \tag{9.10}$$

Fig. 9.8 also shows a 34% and 36% increment in the waste heat performance ratio of the FB-MEE process compared to the conventional MEE consistent with the production rate improvement.

9.3.4 Auxiliary Pumping Power

As discussed in Chapter 5, the pumping power is calculated according to Eq. (9.11).

$$Pumping\ power(\text{kW}) = \left(\frac{\Delta P \cdot \dot{V}}{\eta}\right) \tag{9.11}$$

ΔP, $\dot{V}$, and η are the total head (kPa), volumetric flow rate (m^3/s), and the efficiency of the pump, respectively. As shown in Figs. 9.4 and 9.5, the main pumps of this process are summarized next.

9.3.4.1 Cooling Water Pump

This pump conveys the cooling water through the condenser. The required head for this pump is taken to be 0.4 bar per the number of transfer units (*NTU*) (see Appendix C). A 75% efficiency has been considered for this pump based on onsite specifications [18].

9.3.4.2 Recirculation Pump

This pump is used in the FB-MEE process to recirculate the reconcentrate through the liquid–liquid heat exchanger and the flash vessels. The major duty of this pump is the plate-type heat exchanger pressure drop, which is roughly 0.3 bar per *NTU* (see Appendix C) and the pressure head (see Section 5.2.2). In this case, the required head for this pump is around 88 and 207 kPa for 75°C and 85°C scenario, respectively. According to the site specifications, an 80% efficiency has been considered for this pump [18].

9.3.4.3 Condensate Extraction Pump

This pump is used to extract the condensate from the primary MEE unit. The required head for this pump is considered to be 2 bar (see Section 5.2.4), and the efficiency is considered to be 70% [18].

9.3.4.4 Heat Source Pump

This pump is used for both conventional MEE and FB-MEE processes, and pumps the main heat source medium through the evaporator (first effect) of the primary MEE plant and the liquid–liquid heat exchanger of the FB-MEE process. The required head for this pump is taken to be 0.3 bar per *NTU* for the first effect (evaporator) and for the liquid–liquid heat exchanger in the FB-MEE process (see Appendix C), with a 70% efficiency [18].

9.3.4.5 Reconcentrated Liquor (Feed) Extraction Pump

There are two extraction pumps for the concentrated feed—one to extract from the primary MEE and the other from the flash vessel unit. The required head for this pump is 2 bar (see Sections 5.2.4 and 5.2.5). Based on site specification an 80% efficiency has been considered [18].

9.3.4.6 Feed Pump

The feed pump pumps the feed stream to the evaporation unit. A 1.0 bar head and an 80% efficiency have been considered for this pump [18].

9.3.4.7 NCG Extraction Vacuum Pump

In order to remove the generated noncondensable gasses (*NCGs*) and the air leakage into the system, a water-ring vacuum pump has been considered to remove 1% of the vapor in the condenser as *NCGs* [27]. The corresponding pumping power consumption is evaluated according to the manufacturer's catalogs for water-ring vacuum pumps.

Fig. 9.9 shows the breakdown of the specific pumping power for both systems to highlight the share of each pump in the total specific pumping power. The FB-MEE consumes 5.2

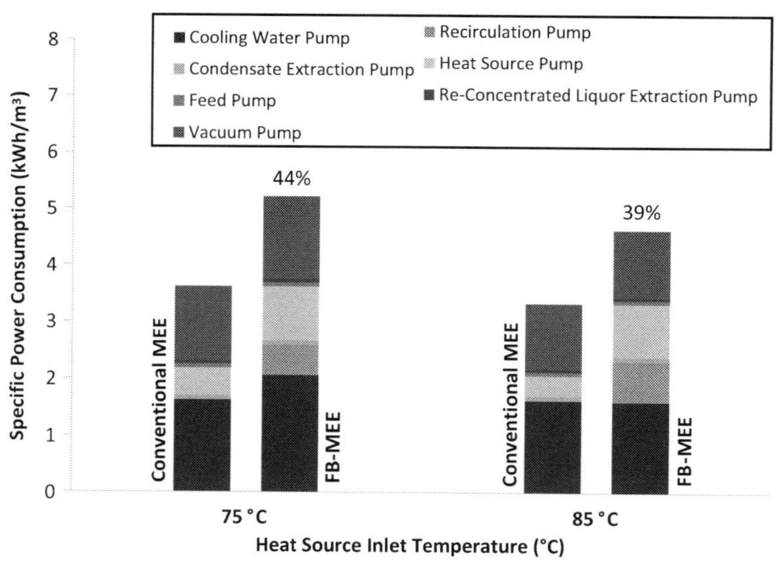

Figure 9.9

Specific pumping power (normalized by the total freshwater production rate) breakdown for both conventional multi-effect evaporation and flash-boosted multi-effect evaporation.

and 4.6 kWh/m^3, which is 44% and 39% more than that of the conventional MEE with 3.6 and 3.3 kWh/m^3 at 75°C and 85°C, respectively. The main difference between the two processes is related to the recirculation pumps deployed in the FB-MEE process. The drop from 5.2 kWh/m^3 at 75°C to 4.6 kWh/m^3 at 85°C is because of the extra primary MEE effect that can be installed into the FB-MEE process on account of the higher heat source inlet temperature as compared to the conventional MEE.

9.3.5 Capital Cost Analysis

The capital costs depend on the material and equipment specifications. For the purpose of comparison, the focus herein is on the relative capital cost increase of the FB-MEE compared to the conventional MEE process and its impact on the specific capital costs. The latter is defined as the ratio of the gross capital cost to the total freshwater production rate, or equivalently to the reconcentrate production rate since the feed-to-vapor ratio is 2.0, which means that there is an equal production of freshwater and reconcentrated process liquor. For conventional thermal desalination plants and the capacity range up to 10,000 m^3/day, the following cost function is used (see Section 7.2):

$$Capital\ Cost_{MEE,MSF}(\text{US\$}) = \Psi_{D_t} = 3054 \times D_t^{0.9751} \tag{9.12}$$

where D_t is the total production rate of the plant in m^3/day and Ψ_{D_t} is the cost function. The capital cost increase is due to the array of flash vessels, the liquid–liquid heat exchanger and the increased heat transfer requirements of the primary MEE effects due to supplementary steam injection as in the FB-MEE process (see Section 7.2). The flash vessels are major parts of a typical MSF process; therefore the cost can be estimated by the cost equation by treating the total vapor produced from the flash chambers as the equivalent MSF plant capacity. However, in contrast to MSF plants, no condenser tubing and no heat rejection section is required for the array of flash vessels, which are also major cost items in thermal desalination plants [16,28]. For the liquid–liquid heat exchanger, based on the type of heat exchanger its price can be calculated. In this case a plate type heat exchanger is considered to be used and its price is estimated according to Appendix E.

As mentioned in Chapter 7, the capital costs for MSF and MEE systems are subdivided into five main sectors [28,29]: capital cost of the evaporator (40% of the total capital cost); equipment piping (29%); erection (14%); engineering and commissioning (10%); and electrical, instrumentation, and control (7%); therefore the total capital cost for the FB-MEE process can be estimated as follows:

$$\begin{aligned} Capital\ Cost_{FBMEE} = {} & \Psi_{D_{t,FBMEE}-D_{FV}} + \left[0.20\cdot\left(\Psi_{D_{t,FBMEE}} - \Psi_{D_{t,FBMEE}-D_{FV}}\right)\right] + 0.41\cdot\Psi_{D_{FV}} \\ & + CC_{hex} \end{aligned} \tag{9.13}$$

Table 9.2: Area coefficients consistent with the degree of superheat in each effect for 85°C heat source inlet temperature processes.

	First Effect (ϕ_1)	Second Effect (ϕ_2)	Third Effect (ϕ_3)	Condenser ($\phi_{cond.}$)
MEE	1	1.08	N/A	1.08
FB-MEE	1	1.08	1.09	1.09

D_{FV} is the production rate (vapor) of all flash vessels and $D_{t,FBMEE}$ is the total production rate of the FB-MEE plant. As mentioned in Section 9.3.2, save the first effect, all vapor introduced to the evaporation effects have around 8°C superheat. This means that extra heat exchanger area for desuperheating in those effects has to be considered, and will give rise to a higher capital cost that impacts on the evaporator section, which is 40% of the total plant capital cost [28,29]. From practical experience [19] 1% extra heating area is needed per 1°C of superheat to account for future scaling problems due to the presence of superheated steam. For this purpose, an area coefficient (ϕ) is applied to address the extra heat exchanger area of the evaporation effects (except for the first effect) as shown in Table 9.2.

$$A_{calc.} = A_1 + A_2 + A_3 + \ldots + A_n + A_{cond.} \tag{9.14}$$

$$A_{actual} = \phi_1 \cdot A_1 + \phi_2 \cdot A_2 + \ldots + \phi_n \cdot A_n + \phi_{cond.} \cdot A_{cond.} \tag{9.15}$$

$$\Upsilon = \frac{A_{actual}}{A_{calc.}} \tag{9.16}$$

$A_{calc.}$ and A_{actual} (in m^2) are the total heat exchanger area of the primary MEE section without and with considering the impact of superheated steam, respectively, and (ϕ) is the area coefficient that should be applied to the evaporator capital cost (40% of the total capital cost); therefore the total capital cost equation is

$$Capital\ Cost(\text{US\$}) = (0.4 \cdot \Upsilon + 0.6) \cdot \Psi_{D_t} \tag{9.17}$$

This correlation should be applied to the three Ψ_{D_t} terms in Eq. (9.13) except for $\Psi_{D_{FV}}$, which relates to flash vessels, and CC_{hex}, which relates to the plate heat exchanger (see Appendix E).

Following this correlation, the FB-MEE scheme incurs around 35% higher total capital costs as compared to a conventional MEE unit (Table 9.3). However, the specific capital cost ratio for both MEE and FB-MEE processes is actually around the same over the application range due to the increased production rate:

$$Specific\ Capital\ Cost\ Ratio = \frac{Capital\ Cost\ Ratio}{Production\ Rate\ Ratio} \tag{9.18}$$

Table 9.3: Capital cost and specific capital cost ratio for flash-boosted multi-effect evaporation as compared with the multi-effect evaporation process.

Heat Source Inlet Temperature	Capital Cost Ratio	Specific Capital Cost Ratio
75°C	1.35	1.01
85°C	1.36	1.00

9.3.6 Conclusion

A novel FB-MEE process has been developed for the exploitation of waste heat streams in an alumina refinery plant to minimize the steam consumption of evaporation units and reduce overall fuel consumption. Its performance has been benchmarked with a conventional MEE process in terms of freshwater and reconcentrated process liquor production and waste–heat performance ratio. The FB-MEE scheme has been demonstrated to be around 36% better than the optimized conventional MEE process with respect to the production rate over the range of waste heat stream temperatures considered. The specific electrical power consumption of this system is 44% and 39% for the inlet heat source temperatures of 75°C and 85°C, respectively. The main reason for the reduction in the pumping power consumption for the 85°C case is because of the extra effect installed in the primary MEE section. A capital cost analysis indicates that these two systems have the same specific capital cost.

9.4 A Novel Flash-Boosted Thermal Vapor Compression Multi-Effect Evaporation for Alumina Refineries

In this section, a novel flash-boosted thermal vapor compression multi-effect evaporation (FB-TVC-MEE) process has been developed. This process utilizes the available waste heat sources of the evaporation units (namely, a low pressure vapor stream) to reduce live steam consumption for evaporation purpose in the alumina refineries.

As mentioned in Section 9.2, evaporation units in alumina refineries typically include tube falling film or flash vessel techniques. Both flashing and evaporation techniques have been shown in Figs. 9.2 and 9.3. As shown in Fig. 9.2, in the MSF configuration, the vapor produced in the last flash vessel is injected into the barometric condenser and is mixed with the cooling water supply stream. In the selected evaporation plant, the inlet temperature of the vapor to the barometric condenser is suitable to run a TVC-MEE process with the aim of saving energy (Fig. 9.10). This section introduces a novel process (Fig. 9.11) that can ultimately be coupled with these available waste heat streams in the evaporation units of an alumina refinery plant to recover the load of a selected evaporation

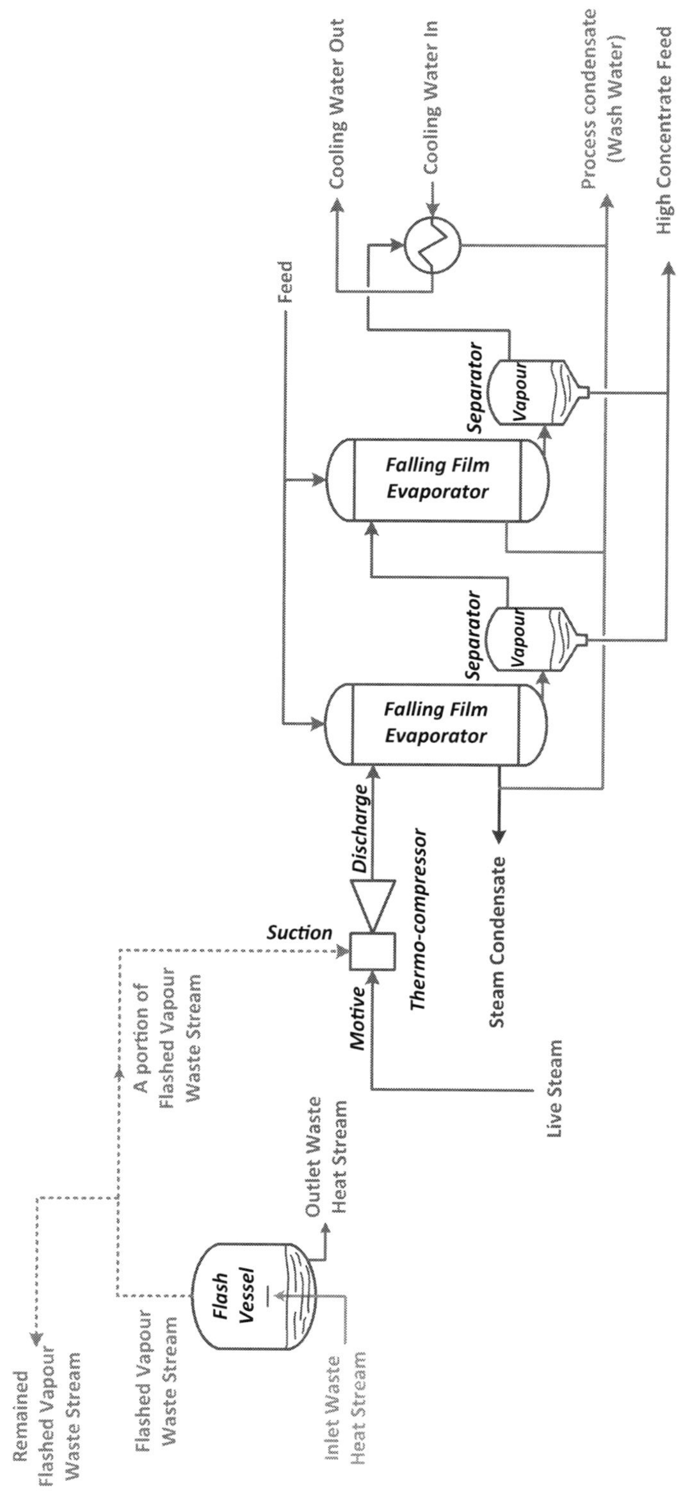

Figure 9.10
Conceptual design of a conventional thermal vapor compression multi-effect evaporation process coupled with the available waste stream from a flash vessel.

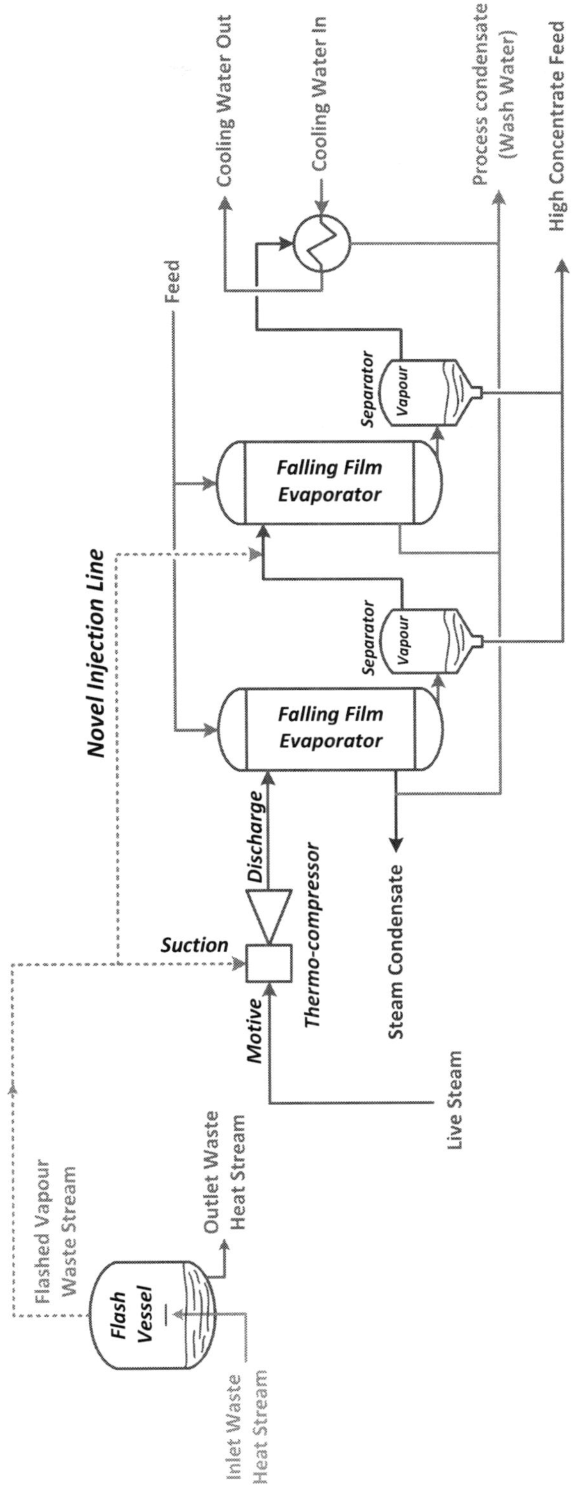

Figure 9.11
Conceptual design of the novel flash-boosted thermal vapor compression multi-effect evaporation process coupled with the available waste stream from a flash vessel.

unit, so as to reduce the live steam consumption (which is the main heat source of the evaporation process) as compared to both the current MSF process and the conventional TVC-MEE process.

9.4.1 Process Description

The selected evaporation plant includes some evaporation units, which uses the flashing technique. The purpose of the novel design is to utilize the available flashed vapor waste stream to recover the load of a selected evaporation unit and eliminate its live steam consumption. Table 9.4 summarizes the selected evaporation units' specifications. However, in view of commercial confidentiality, the quantities that are shown are different from the actual specifications, but are meant to be generic and representative. For this purpose the simulation has been done for unit mass flow rate of live steam. Also the feed liquor stream has been replaced with pure caustic soda (NaOH) solution, which is more conservative due to its higher *BPE* as compared to the actual feed liquor stream.

As shown in Table 9.4, the selected MSF evaporation unit is considered to consume 1 kg/s live steam to reconcentrate 18.45 kg/s of 19.0 wt% caustic soda feed stream to 16.30 kg/s of 21.5 wt% caustic soda. As shown in Fig. 9.2 the considered available waste stream is the flashed vapor from the last flash vessel of the other evaporation unit. This stream has enough potential to recover the load of the selected unit. For benchmarking purpose an optimized conventional TVC-MEE process as shown in Fig. 9.12 is considered to harness the available waste heat source. Based on our process simulation, which is detailed later, it consumes 35% of the available live steam (0.35 kg/s as opposed to the original 1 kg/s consumption) via a thermo-compressor to compress around 56% of the aforementioned available flashed vapor in order to recover the duty of the selected evaporation unit. The remaining flashed vapor (44%) from the aforementioned flashed vapor stream is then

Table 9.4: Assumed specifications.

Live Steam	Pressure (bara)	5
	Temperature (°C)	170
	Flow rate (kg/s)	1.00
Feed inlet (caustic soda)	Concentration (wt%)	19.0
	Temperature (°C)	59.0
	Flow rate (kg/s)	18.45
Equivalent process liquor outlet (concentrated caustic soda)	Concentration (wt%)	21.5
	Flow rate (kg/s)	16.30
Available waste heat stream (flashed vapor)	Temperature (°C)	72.0
	Pressure (bara)	0.25
	Flow rate (kg/s)	1.55
Cooling water supply	Temperature (°C)	33.0

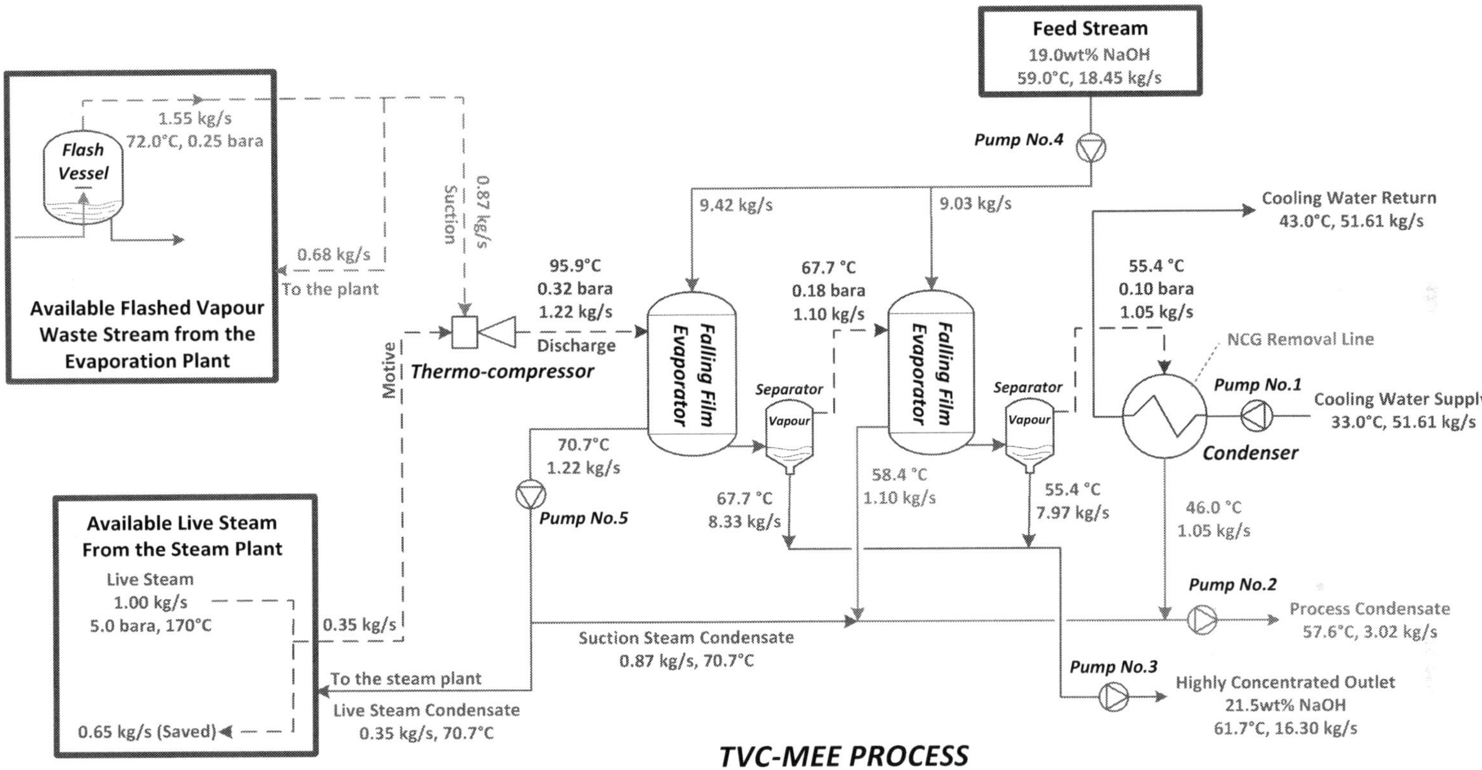

Figure 9.12

Schematic design of the conventional thermal vapor compression multi-effect evaporation process coupled with the available waste stream (as the benchmark) of the considered alumina refinery plant. Other than plant specifications, all figures are based on our simulation.

purged as a waste stream to the extant barometric condenser. This configuration does not have any potential to use the remaining waste stream. In case of using more flashed vapor, the discharge saturated temperature will decrease and result in a reduction in the production rate. The process parameters in Fig. 9.12, other than those specifications as in Table 9.4, are the results of our process simulation.

In contrast, as shown in Fig. 9.13, the FB-TVC-MEE process is used to harvest the same flashed vapor stream. In this process, our simulation as detailed in the following section indicates that only around 18% of the selected unit live steam (0.18 kg/s as opposed to the original 1 kg/s consumption) is needed to compress 28% of the available flashed vapor stream. The remaining flashed vapor (72%) is then summarily injected into the second effect of the MEE system. This augmentative steam injection process has been elucidated in our previous works known respectively as the boosted MEE (B-MEE) and flash-boosted MEE (FB-MEE) processes [11,16–18,26,30–34]. Essentially this is a special application of our earlier reported FB-MEE scheme that is coupled to the thermo-compressor, whereby the primary MEE process is limited to two effects, and the battery of flash chambers is reduced to one flash chamber. Again the process parameters in Fig. 9.13, other than those specifications as in Table 9.4, are the results of our simulation.

The virtue of flash vapor injection in the FB-TVC-MEE process is to minimize the suction load of the thermo-compressor, which then economizes on live steam consumption. As shown in Figs. 9.12 and 9.13, the live steam consumption of the novel FB-TVC-MEE process is 49% less than the optimized conventional TVC-MEE process (0.18 kg/s as compared with 0.35 kg/s live steam consumption). Due to the higher flashed vapor recovery of the FB-TVC-MEE process as compared to the conventional TVC-MEE process, the process condensate (wash water) production rate increases by 23%.

This novel FB-TVC-MEE process will be compared with the conventional TVC-MEE process and the current MSF process, in terms of the amount of saved live steam, gain output ratio (*GOR*), specific pumping power consumption, and specific capital cost.

9.4.2 Technical Simulation

Our simulations encompass mass, concentration, and energy balances for the MEE effects, the condenser, and the thermocompressor. The GRG method [35] is used to solve the accompanying equations by imposing boundary conditions stemming from the law of thermodynamics and appropriate operational, technical, and economic constraints as shown in Table 9.5. Validated simulations and their methodologies have been reported in detail in our previous works [11,16,17,30,32–34]. The basic simulation has been validated

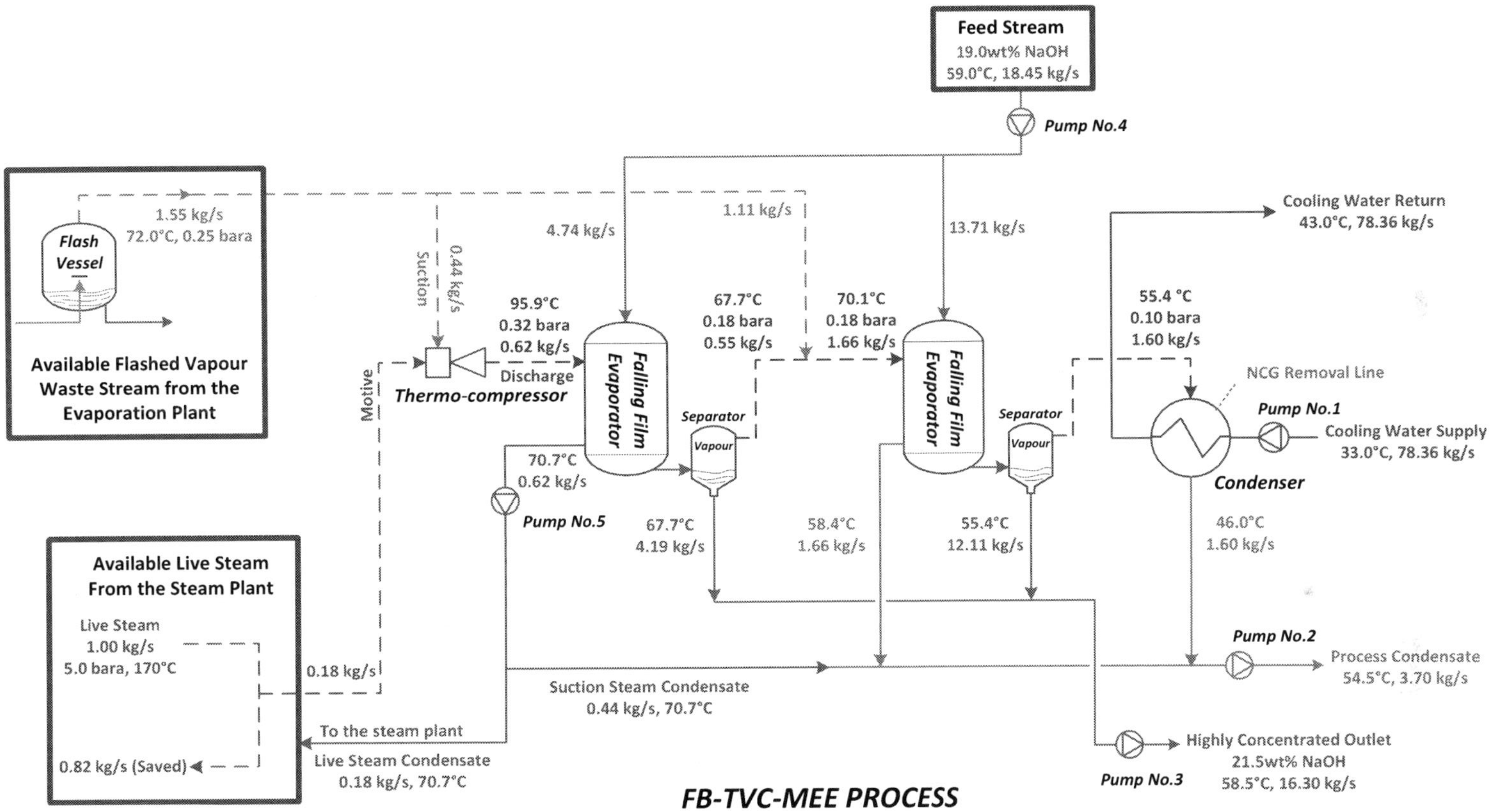

Figure 9.13

Schematic design of the flash-boosted thermal vapor compression multi-effect evaporation process coupled with the available waste stream of the considered alumina refinery plant. Other than plant specifications, all figures are based on our simulations.

Table 9.5: Assumptions and boundary conditions of our thermodynamic simulation.

Recovery factor for each multi-effect evaporation effect ($1/R$)	12%
Temperature difference across the condenser (°C)	10.0
Minimum pressure difference for steam injection purpose (kPa)	1.0 [33]
Temperature of approach across the evaporators (°C)	3.0
Feed (NaOH) inlet/outlet concentration (wt%)	19.0/21.5
Maximum available waste flashed vapor mass flow rate (kg/s)	1.55
Maximum available live steam mass flow rate used as the motive steam in the thermocompressor (kg/s)	1.00

against market available data from a reputable manufacturer [17,36], and our pilot plant [11] (see Chapter 3).

Due to the high *BPE* of NaOH as the feed liquor, which is 9.4°C for pure 21.5 wt% NaOH, and the optimized saturated temperature difference between the steam after vapor compression and the cooling water supply (heat sink), the number of effects in the MEE process is limited to two effects (Figs. 9.12 and 9.13). The feed liquor (NaOH) enters the MEE effects at 19.0 wt% and 58.4°C and is concentrated to 21.5 wt%, being identical to the extant inlet and outlet conditions of the selected evaporation unit for process liquor. A low recovery factor (12%) is chosen to match the extant flashing process so as to harmonize with the entire operation. We note however that the potential of an MEE process, in terms of recovery factor, is much more than that [12].

The simulation aims to find the lowest possible required live steam mass flow rate (to wit, the motive steam of the thermocompressor) and to calculate the required suction flow rate of the thermo-compressor, and the injected vapor mass flow rate in order to have the same concentrate production rate as the selected evaporation unit. It is also calculated that 70.7°C saturated steam temperature is the corresponding optimized heat source inlet temperature to have a two-effect MEE to meet the required concentrate production rate.

The mass and concentration balances are expressed as $k \in \{1, 2\}$:

$$\dot{m}_{F,k} = \dot{m}_{HC,k} + \dot{m}_{V,k} \tag{9.19}$$

$$\dot{m}_{F,k}\cdot X_{F,k} = \dot{m}_{HC,k}\cdot X_{HC,k} \tag{9.20}$$

for which

$$\dot{m}_{HC,k} = \left(\frac{R-1}{R}\right)\cdot \dot{m}_{F,k} \tag{9.21}$$

$$X_{HC,k} = \left(\frac{R}{R-1}\right)\cdot X_{F,k} \tag{9.22}$$

The energy balance for the first effect is written as

$$\dot{m}_{HS,1}\cdot\left(h_{HS,1,in} - h_{f_{\langle P_d\rangle}}\right) = \dot{m}_{V,1}\cdot h_{V,1} + \dot{m}_{HC,1}\cdot h_{HC,1} - \dot{m}_{F,1}\cdot h_{F,1} = U_1\cdot A_1\cdot \Delta T_{lm,1} \tag{9.23}$$

where $h_{f_{\langle P_d\rangle}}$ is the saturated enthalpy of the discharge steam condensate upon leaving the first effect; $h_{HS,1,in} = h_{g_{\langle P_d,T_d\rangle}}$ is the enthalpy of superheated steam at discharge pressure and temperature; $h_{V,1} = h_{g_{\langle P_1,T_{HC,1}\rangle}}$ is the superheated vapor at the effect's pressure and the outlet concentrate temperature; $h_{HC,1}$ and $h_{F,1}$ are the enthalpies of the concentrate outlet and feed inlet streams at their relevant concentration and temperature, respectively.

The energy balance for the second effect, which receives waste flash vapor by injection, in the case of FB-TVC-MEE, is expressed as

$$\dot{m}_{HS,2}\cdot\left(h_{HS,2,in} - h_{f_{\langle P_1\rangle}}\right) = \dot{m}_{V,2}\cdot h_{V,2} + \dot{m}_{HC,2}\cdot h_{HC,2} - \dot{m}_{F,2}\cdot h_{F,2} = U_2\cdot A_2\cdot \Delta T_{lm,2} \tag{9.24}$$

For which,

$$\dot{m}_{HS,2} = \dot{m}_{V,1} + \theta\cdot \dot{m}_{V,inj} \tag{9.25}$$

and

$$h_{HS,2,in} = \frac{\dot{m}_{V,1}\cdot h_{V,1} + \theta\cdot \dot{m}_{V,inj}\cdot h_{V,inj}}{\dot{m}_{V,1} + \theta\cdot \dot{m}_{V,inj}} \tag{9.26}$$

$$h_{V,2} = h_{g_{\langle P_2,T_{HC,2}\rangle}} \tag{9.27}$$

$$h_{HC,2} = h_{f,HC_{\langle X_{HC,2},T_{HC,2}\rangle}} \tag{9.28}$$

$$h_{F,2} = h_{f,F_{\langle X_{F,2},T_{F,2}\rangle}} \tag{9.29}$$

θ is a flag that distinguishes TVC-MEE from the FB-TVC-MEE process, so that $(\theta = 0)$ for TVC-MEE and $(\theta = 1)$ for FB-TVC-MEE.

The energy balance for the condenser is written as

$$\dot{m}_{V,2}\cdot\left(h_{V,2} - h_{f_{\langle P_2\rangle}}\right) = \dot{m}_C\cdot\left(h_{C,out} - h_{C,in}\right) = U_{cond.}\cdot A_{cond.}\cdot \Delta T_{lm,cond.} \tag{9.30}$$

where $h_{C,in}$ and $h_{C,out}$ are the enthalpies of the inlet and outlet cooling water at their relevant temperatures, respectively.

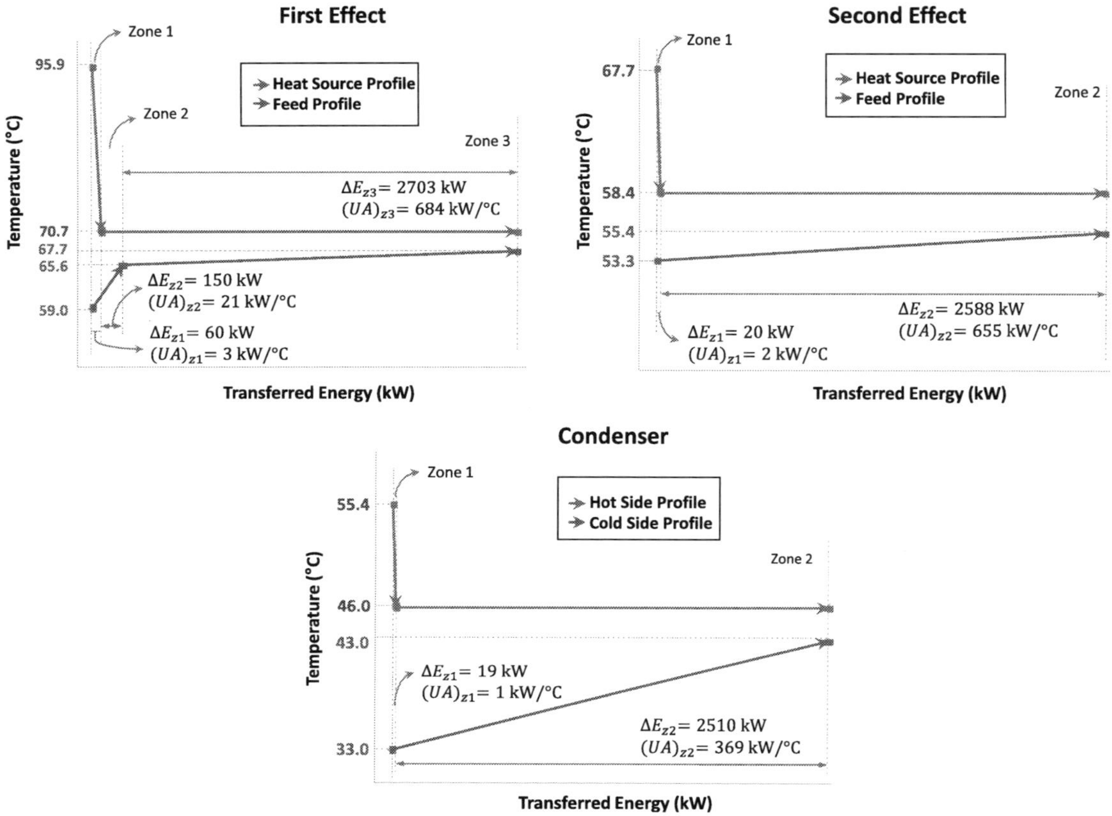

Figure 9.14

Schematic of the temperature–energy profiles for the first effect, second effect, and the condenser of the conventional thermal vapor compression multi-effect evaporation (TVC-MEE). All figures are based on our simulation.

Figs. 9.14 and 9.15 illustrate the temperature–energy profiles across the first effect, second effect, and condenser for the conventional TVC-MEE and the FB-TVC-MEE processes, with the values stemming from our simulation. The heat source of these effects and the condenser is the superheated vapor from the previous effect or the discharge steam from the thermocompressor. Referring to the first effect (Figs. 9.14 and 9.15), in zone 1 the superheated vapor first desuperheats and then condenses over zones 2 and 3. In contrast, the feed temperature increases to the relevant boiling temperature and remains constant over the evaporation process. In this effect (effect 1), there are three different zones with three different modes of heat transfer. In zone 3, the feed is boiling and therefore its concentration is changing from 19.0% to 21.5% (Table 9.5) and because the *BPE* is a function of the concentration, it is changing during the evaporation process, so that this process as shown in Figs. 9.14 and 9.15 is not a constant temperature evaporation process. The *UA* values for zones 1 and 2 are not more than 8% of the total *UA* value in this effect for the FB-TVC-MEE process (3% for the conventional TVC-MEE process), and therefore

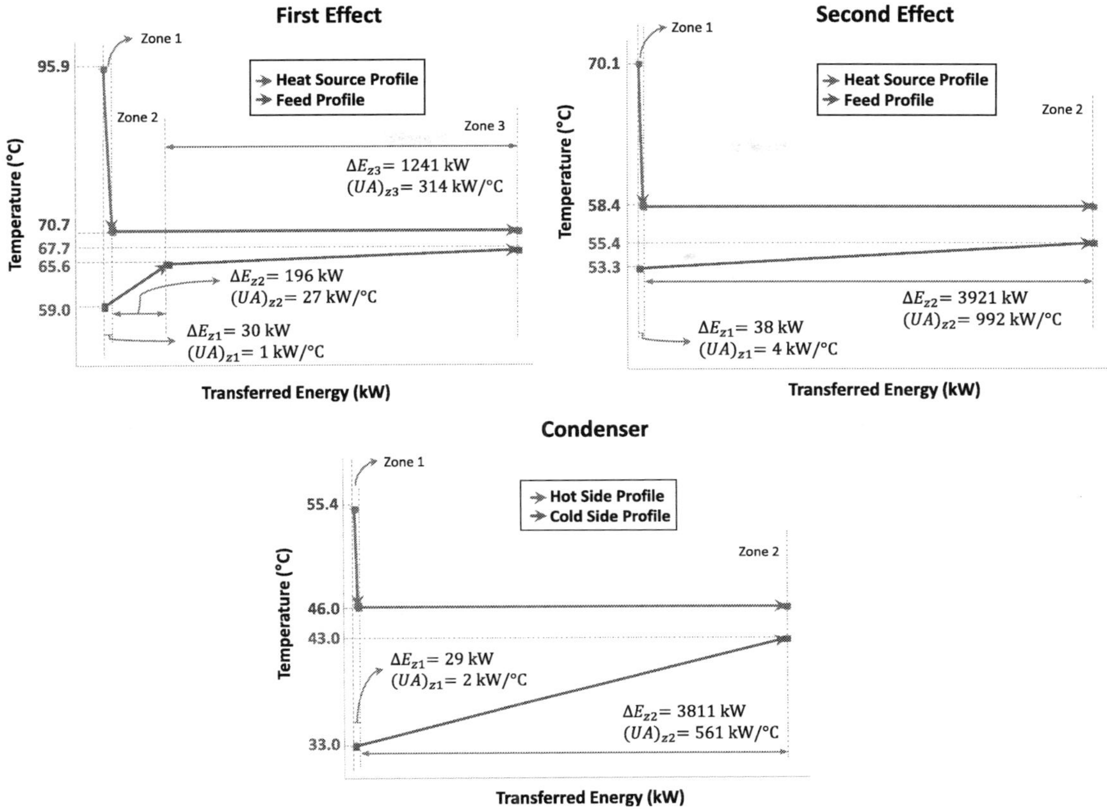

Figure 9.15

Schematic of the temperature–energy profiles for the first effect, second effect, and the condenser of the flash-boosted thermal vapor compression multi-effect evaporation. All figures are based on our simulation.

to a good approximation their relevant heat transfer coefficients can be ignored, so that the overall U value for this effect (U_1) is simply U_{z3}, which pertains to zone 3, where latent–latent heat transfer prevails (see Appendix D).

In the second effect, there are two zones. In zone 1, the superheated vapor first desuperheats and then condenses over zone 2. In our simulation for the second effect the impact of flashed feed stream is ignored. Flashing happens when the feed saturation pressure is higher than the pressure of the second effect. Hence a negligible amount of feed (less than 1%, that commensurate with 5.7°C of sensible heat in our case), is flashed while its pressure and temperature decrease (namely, from 59°C to 53.3°C). For the second effect in both the FB-TVC-MEE and TVC-MEE the inlet feed temperature is considered to be 53.3°C, namely the relevant feed boiling point temperature (which is commensurate with its concentration) that corresponds to the pressure in the second effect (Figs. 9.14 and 9.15).

After flashing the feed is boiling while its concentration is changing from 19.0% to 21.5% (Table 9.5). As mentioned earlier, due to the different *BPE*, the temperature varies during the evaporation process. Again, as shown in Figs. 9.14 and 9.15, to a very good approximation the overall U value for this effect (U_2) is simply U_{z2}.

Referring to Figs. 9.14 and 9.15, the cooling water temperature increases across the condenser but the trend of the superheated vapor temperature profile is the same as in the previous effects, so that zone 1 (the desuperheating zone) is negligible once again, with the overall heat transfer coefficient ($U_{cond.}$) being practically U_{z2}, where latent-sensible heat transfer happens (see Appendix D).

In our simulation as a conservative assumption a 3°C temperature difference has been considered between the concentrate outlet boiling temperature and the heat source condensation temperature.

For the thermo-compressor, the required motive steam and produced discharge steam are calculated according to Fig. 9.16 [37], with the assumed suction (P_s), motive (P_m), and discharge (P_d) pressures, and the required expansion ($EXP = P_m/P_s$), compression ($CMP = P_d/P_s$), and entrainment (ENT) ratios. The next two equations are used to calculate the required motive and suction mass flow rates.

$$\dot{m}_m = \dot{m}_s / ENT \tag{9.31}$$

where ENT is the entrainment ratio to be found from Fig. 9.16 [37] and

$$\dot{m}_d = \dot{m}_s + \dot{m}_m \tag{9.32}$$

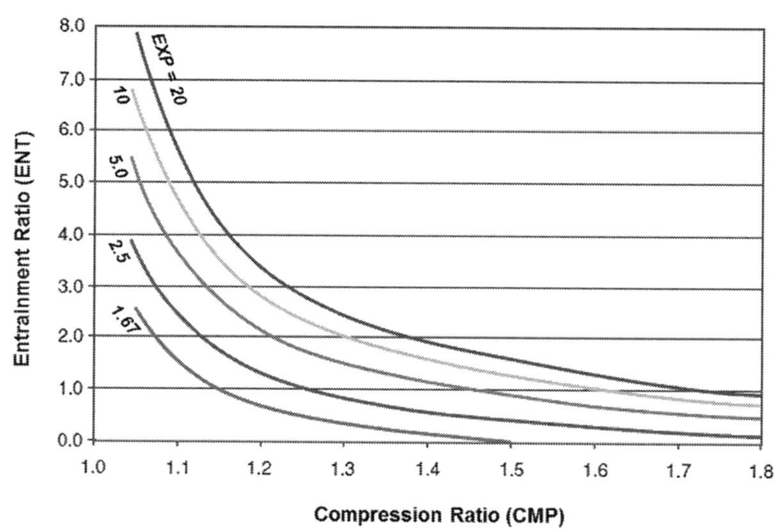

Figure 9.16
Thermocompressor entrainment chart [37].

The discharge enthalpy and temperature can be found from the energy balance equation, which is written as

$$h_d = \frac{\dot{m}_m \cdot h_m + \dot{m}_s \cdot h_s}{\dot{m}_s + \dot{m}_m} \tag{9.33}$$

where $h_m = h_{g\langle P_m, T_m \rangle}$ and $h_s = h_{g\langle P_s, T_s \rangle}$. The discharge temperature $T_d = f(h_d, P_d)$, which is a function of the discharge pressure and enthalpy, can be obtained from steam property routine. This temperature shows the degree of superheating that enjoins extra heat exchanger area to cater for desuperheating.

The overall amount of wash water generated is the sum total of the condensate flow rate from each of the MEE effects:

$$\dot{M}_{D,total} = \sum_{k=1}^{2} \dot{m}_{V,k} + \theta \cdot \dot{m}_{V,inj} + \dot{m}_s \tag{9.34}$$

$$\dot{M}_{HC,total} = \sum_{k=1}^{2} \dot{m}_{HC,k} \tag{9.35}$$

As mentioned earlier, the GRG method is used to solve Eqs. (9.19)–(9.35) to find the lowest possible motive steam flow rate to produce the same concentrate flow rate as the conventional MSF process (the original process) by imposing the boundary conditions for both the conventional TVC-MEE and FB-TVC-MEE processes. In all the previous equations, and for the properties of steam and vapor, the REFPROP package (REFerence fluid PROPerties) as developed by the National Institute of Standards and Technology [38] has been used. The properties of NaOH are consulted from [20–24,39,40]. Extensive details related to the simulation of the MEE process can be found in Refs. [12,16].

9.4.3 Capital Cost Analysis

The overall capital cost is a key parameter for the financial decision makers. For conventional thermal desalination plants the capital cost can be approximated based on the GWI/IDA database [41]. For the capacity range up to 10,000 m^3/day, the following conservative cost function applies [25]:

$$\Psi_{D_t} = 3054 \times D_t^{0.9751} \tag{9.36}$$

where D_t is the total distillate production rate of the plant in (m^3/day) and Ψ_{D_t} is the conservative cost function that yields the total capital cost of the relevant plant. The capital costs for thermal distillation processes (MSF and MEE systems) can be divided into two main sections [28,29]: capital cost of the evaporator (40% of the total capital cost); and remaining items that include equipment piping, erection, engineering, and commissioning as well as electrical, instrumentation, and control (60%). While this cost function stems from seawater applications with different choice of material compared with process liquor operation (i.e., the use of cast iron), since we are only interested in percentage cost differential comparison between two thermal distillation processes, the comparative procedure remains valid.

Eq. (9.36) is used to calculate the capital cost of TVC-MEE as a function of process condensate (distillate) production rate, but for FB-TVC-MEE because of the injected flashed vapor to the second effect, its impact on the capital cost should be evaluated. This procedure has been developed in our previous work [16,25]. Accordingly, the capital cost of an FB-TVC-MEE system can be analyzed with the help of the full cost function of Eq. (9.36) by considering the process condensate capacity of the FB-TVC-MEE system, save the injected flashed vapor flow rate to the second effect. For the vapor injected from the waste flashed vapor stream, only condensation takes place in the MEE section and no additional evaporation surface needs to be considered for the vapor generation in this section, insofar as capital cost analysis is concerned. By assuming similar overall heat transfer coefficients for both evaporation and condensation processes [12], only half of the typical heat transfer area is therefore needed to process the additional vapor injection, which is then translated to a 50% discount of the earlier introduced cost breakdown for the evaporator (50% of 40% for the cost of evaporator) [16,25]. This overall factor is then applied to the capital cost differential between an MEE plant with a process condensate production capacity of an FB-TVC-MEE plant and an MEE plant with a process condensate production capacity of an FB-MEE plant less the total vapor injection rate from the flashed vapor waste stream as shown in Eq. (9.37) [16,25].

$$CC_{\langle FB-TVC-MEE\rangle} = \Psi_{\left(D_{t\langle FB-TVC-MEE\rangle} - D_{inj.}\right)} + \left[0.2\cdot\left(\Psi_{D_{t\langle FB-TVC-MEE\rangle}} - \Psi_{\left(D_{t\langle FB-TVC-MEE\rangle} - D_{inj.}\right)}\right)\right] \tag{9.37}$$

where $D_{t\langle FB-TVC-MEE\rangle}$ is the total process condensate (distillate) flow rate (m^3/day) of the FB-TVC-MEE and $D_{inj.}$ is the injected flashed vapor flow rate (m^3/day) into the second effect of the MEE section.

The discharge steam from the thermo-compressor, and for that matter, all the vapors (namely injected flashed vapor and vapor produced from the MEE process) has an inherent high degree of superheat. Consequently extra heat exchanger area for desuperheating in those evaporation effects has to be considered, and gives rise to a higher capital cost that impact on the evaporator section, which constitutes 40% of the total plant capital cost [28,29]. From practical experience [19], 1% extra heating area is needed per 1°C of superheat in order to account for both desuperheating and future scaling problems. For this purpose, according to the degree of superheat, a corresponding coefficient factor is therefore applied to cover the extra heat exchanger area of the evaporation effects as follows:

$$A_{calc.} = A_1 + A_2 + A_{cond} \tag{9.38}$$

$$A_{actual} = \phi_1\cdot A_1 + \phi_2\cdot A_2 + \phi_{cond}\cdot A_{cond} \tag{9.39}$$

and

$$\Upsilon = \frac{A_{actual}}{A_{calc.}} \tag{9.40}$$

where ϕ is the local area coefficient factor, which is considered as the extra percentage of required area per degree of superheat. $A_{calc.}$ and A_{actual} (m^2) are the total heat exchanger areas of the MEE section with and without considering the impact of superheated steam, respectively, and Υ is the overall area coefficient factor that should be applied to the capital cost of evaporator (40% of the total capital cost), and so the resultant total capital cost equation is as follows:

$$TCC(\text{US\$}) = (0.4 \cdot \Upsilon + 0.6) \cdot (CC) \tag{9.41}$$

where CC can be the capital cost for either TVC-MEE process, Eq. (9.36), or FB-TVC-MEE process, Eq. (9.37). The specific capital cost can then be calculated by Eq. (9.42):

$$SCC\left(\text{US\$}/\left(\text{m}^3/\text{day}\right)\right) = \frac{TCC}{D_t} \tag{9.42}$$

where D_t is the process condensate (distillate) production rate (m^3/day).

9.4.4 Simulation Results

9.4.4.1 Steam Consumption, Process Optimization, and Gain Output Ratio

Based on all the previously mentioned methods and assumptions, Table 9.6 and Fig. 9.17 illustrate the amount of steam saving of the two proposed processes as compared to the current process. Evidently, the FB-TVC-MEE process is superior with around 82% saving

Table 9.6: Available live steam, steam consumptions and recovered flashed vapor for the considered processes.

	Selected Evaporation Unit (*MSF*)	*TVC-MEE*	*FB-TVC-MEE*
Available live steam (for the selected evaporation unit) (kg/s)	1.00	1.00	1.00
Live steam consumption (kg/s)	1.00	0.35	0.18
Available flashed vapor waste stream (kg/s)	1.55	1.55	1.55
Recovered flashed vapor waste stream (kg/s)	0	0.87	1.55

FB-TVC-MEE, flash-boosted thermal vapor compression multi-effect evaporation; *MSF*, multi-stage flash; *TVC-MEE*, thermal vapor compression multi-effect evaporation.

of the available live steam and its ability to recover all available flashed vapor waste streams.

For the optimized TVC-MEE process in Fig. 9.12, there is no steam injection into the second effect and consequently around 44% of the flashed vapor waste stream remains unused (0.68 kg/s out of 1.55 kg/s). Fig. 9.13 shows an optimized FB-TVC-MEE process that fully consumes all the flashed vapor waste stream, whereby 28% of which is subjected to the thermo-compressor and the balance is summarily injected into the second effect. Fig. 9.18 elucidates the optimization pathway progressing from an optimized TVC-MEE to an optimized FB-TVC-MEE. Verily the live steam consumption (motive steam flow rate) decreases linearly from 0.35 kg/s for the optimized TVC-MEE (which corresponds to nil injected flashed vapor flow rate) through to 0.18 kg/s for the optimized FB-TVC-MEE process (at 1.11 kg/s injected flashed vapor flow rate). Concomitantly the amount of unused flashed vapor waste steam decreases linearly to zero as the process evolves from an optimized TVC-MEE to an optimized FB-TVC-MEE.

The *GOR* is a key factor, which compares the thermal performance of all three processes. This factor is defined as the amount of favorable production produced by one unit of consumed live steam.

$$GOR = \frac{\dot{m}_p}{\dot{m}_m} \tag{9.43}$$

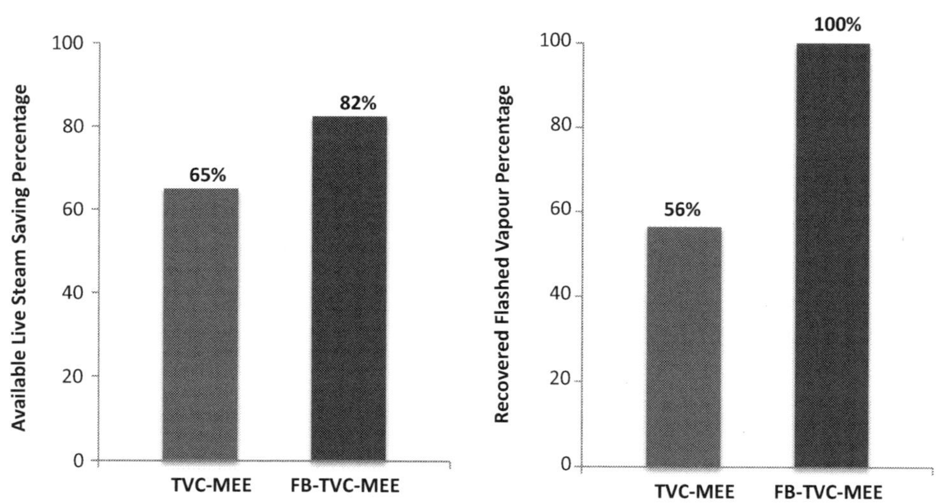

Figure 9.17
Live steam saving and percentage recovery of flashed vapor waste stream for thermal vapor compression multi-effect evaporation (TVC-MEE) and flash-boosted thermal vapor compression multi-effect evaporation processes (FB-TVC-MEE), benchmarked against the extant multi-stage flash process.

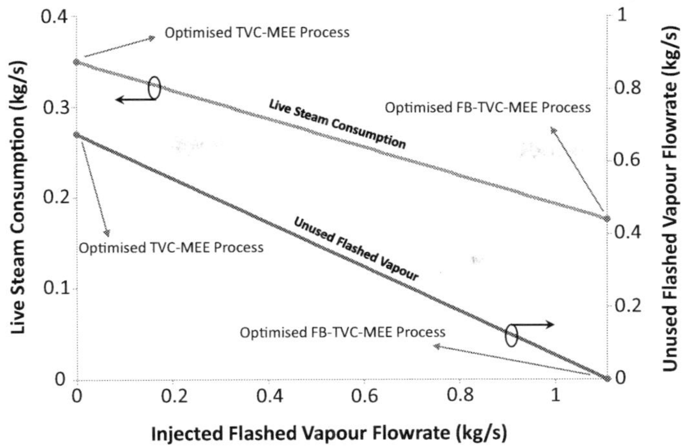

Figure 9.18

Optimization trajectory as the process evolves from an optimized thermal vapor compression multi-effect evaporation (TVC-MEE) to an optimized flash-boosted thermal vapor compression multi-effect evaporation (FB-TVC-MEE).

where $\dot{m}_m$ is the motive steam flow rate and $\dot{m}_p$ is the total produced process condensate excluding the injected vapor from the waste stream, as the latter has always been available. Fig. 9.19 shows the *GOR* for all three considered evaporation processes. Verily, the *GOR* of the FB-TVC-MEE is the highest and around 5.7 times more than the selected evaporation unit and 98% better than the TVC-MEE process, which means it can utilize the live steam more effectively.

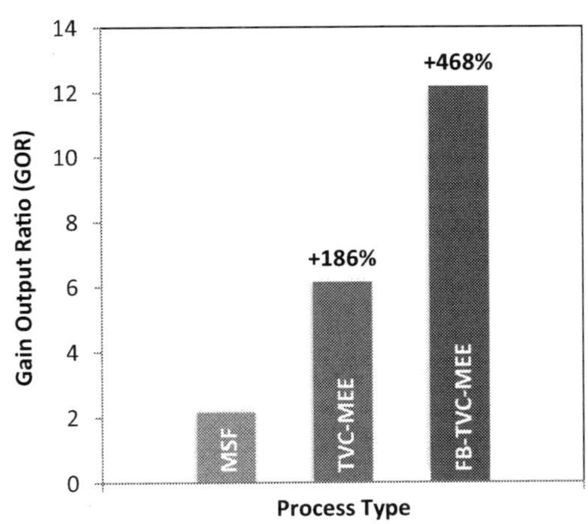

Figure 9.19

Gain output ratio of the current multi-stage flash (MSF), thermal vapor compression multi-effect evaporation (TVC-MEE), and flash-boosted thermal vapor compression multi-effect evaporation (FB-TVC-MEE) processes.

9.4.4.2 Pumping Power Consumption

With respect to pumping power consumption, as indicated in Figs. 9.12 and 9.13, there are five essential pumps to supply the cooling water (pump no. 1), extract the process condensate (pump no. 2) and concentrate (pump no. 3) from the MEE effects, feed the effects (pump no. 4), and extract steam condensate from the first effect (pump no. 5). The pumping power is calculated according to Eq. (9.44):

$$Pumping\ power(\text{kW}) = \left(\frac{\Delta P \cdot \dot{V}}{\eta}\right) \tag{9.44}$$

where ΔP, $\dot{V}$ and η are the total head (kPa), volumetric flow rate (m^3/s), and the overall efficiency of the pump, respectively. Table 9.7 includes the considered pressure heads and efficiencies for the relevant aforementioned pumps. The specific pumping power of FB-TVC-MEE is compared with TVC-MEE as in Fig. 9.20. A detailed calculation of power consumption and its methodology has been explained in Chapter 5 as well. As shown in Fig. 9.20, the main duty is related to the cooling water supply pump (pump no. 1). With respect to the process condensate production rate, due to the higher flashed vapor recovery in FB-TVC-MEE as compared to the TVC-MEE process, the specific pumping power is only 5% more than that of TVC-MEE.

9.4.4.3 Specific Capital Cost

Following the aforementioned method for comparing the capital costs of both TVC-MEE and FB-TVC-MEE processes as in Section 9.4.3, Table 9.8 delineates the extent of superheating in each effect of both processes, which influence both the local evaporator (ϕ) and overall area (U) coefficient factors. Accordingly, the local evaporator coefficients (ϕ) for the first effect, second effect, and the condenser of the TVC-MEE process are 1.252, 1.094, and 1.094, respectively. As for the FB-TVC-MEE, they are 1.252, 1.118, and 1.094, respectively. By applying these factors to Eq. (9.39) and then Eq. (9.40) the overall area coefficient factors (Υ) are 1.16 for the TVC-MEE and 1.14 for the FB-TVC-MEE, and with which the total and specific capital costs can be evaluated according to

Table 9.7: Pump specifications.

Pump No.	Duty	Overall Efficiency	Pressure Head (barg)
1	Cooling water intake pump	0.70	5.1
2	Process condensate extraction pump	0.70	8.5
3	High concentrate liquor extraction pump	0.63	12.3
4	Feed pump	0.77	1.0
5	Steam condensate pump	0.60	12.2

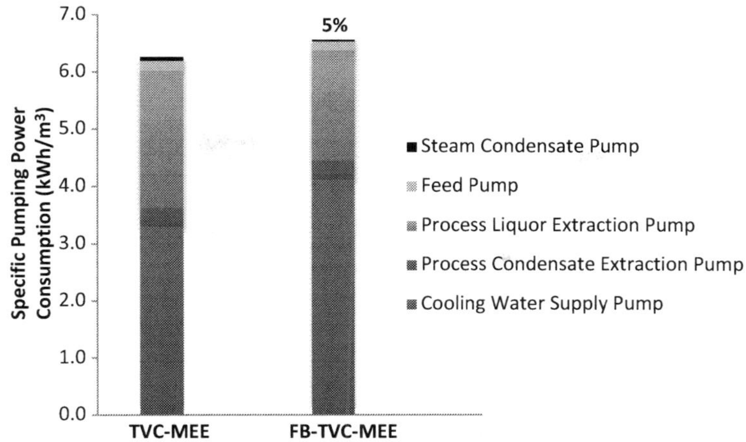

Figure 9.20
Specific pumping powers for the two proposed evaporation processes.

Eqs. (9.41) and (9.42). Based on this procedure, and on account of the higher recovery of the flashed vapor waste stream in the FB-TVC-MEE as compared to the TVC-MEE process, the specific capital cost is around 11% less than that of the TVC-MEE process.

9.4.5 Conclusion

A novel FB-TVC-MEE process has been developed for the exploitation of the available waste heat streams in an alumina refinery plant. This system is meant to be coupled to the available flashed vapor waste streams, so as to recover the load of a selected evaporation unit. Its performance has been benchmarked with an optimized conventional TVC-MEE system and also the current selected unit in terms of process condensate (wash water), *GOR*, specific pumping powers, and specific capital costs reduction. This novel process can save around 82% of the selected evaporation unit's available live steam. Its thermal performance is 468% better than the current system and 98% better than the TVC-MEE process. With respect to pumping power consumption, it shows a 5% increment for the FB-TVC-MEE as compared to the TVC-MEE process. A capital cost analysis indicates an 11% decrement for the overall specific capital cost, for the novel FB-TVC-MEE process

Table 9.8: Degree of superheat in each effect.

	First Effect (°C)	Second Effect (°C)	Condenser (°C)
TVC-MEE	25.2	9.4	9.4
FB-TVC-MEE	25.2	11.8	9.4

FB-TVC-MEE, flash-boosted thermal vapor compression multi-effect evaporation; *TVC-MEE*, thermal vapor compression multi-effect evaporation.

as compared to the TVC-MEE process. Importantly FB-TVC-MEE saves 26% more live steam than the optimized conventional TVC-MEE, which makes this novel process a superior option, in terms of significantly slashing the operating costs and emission of the plant.

References

[1] International Aluminium Institute, Aluminium and Durability towards Sustainable Cities, Cwningen Press, 2014.
[2] T.J. Brown, L.E. Hetherington, S.D. Hannis, T. Bide, A.J. Benham, N.E. Idoine, P.A.J. Lusty, World Mineral Production 2003–2007, Natural Environment Research Council (NERC), 2007.
[3] G. McNamara, Iron fact-ite, Earth Sci. Soc. (2000).
[4] Australian Aluminium Council Ltd, Sustainability Report 2012, Australian Aluminium Council Ltd, 2012.
[5] M. Grafe, G. Power, C. Klaubus, Review of Current Bauxite Residue Management, Disposal and Storage: Practices, Engineering and Science, CSIRO, Karawara WA 6152, Australia, 2009.
[6] G.S. Gontijo, A.C.B. De Araújo, S. Prasad, L.G.S. Vasconcelos, J.J.N. Alves, R.P. Brito, Improving the Bayer process productivity – an industrial case study, Miner. Eng. 22 (13) (2009) 1130–1136.
[7] G. Power, M. Gräfe, C. Klauber, Bauxite residue issues: I. Current management, disposal and storage practices, Hydrometallurgy 108 (1–2) (2011) 33–45.
[8] L. Liu, L. Aye, Z. Lu, P. Zhang, Analysis of the overall energy intensity of alumina refinery process using unit process energy intensity and product ratio method, Energy 31 (2006) 1167–1176.
[9] Q. Chen, Discussion on the ways to reduce steam consumption in alumina production, J. Energy Sav. Nonferrous Metall. 1 (2006) 27–29.
[10] L. Jiang, Discussion of application of plate type evaporator in Shanxi alumina plant, J. Liaoning Chem. Ind. 33 (6) (2004) 345–348.
[11] A. Christ, K. Regenauer-Lieb, H.T. Chua, Application of the Boosted MED process for low-grade heat sources—A pilot plant, Desalination 366 (2015) 47–58.
[12] H.T. El-Dessouky, H.M. Ettouney, Fundamentals of Salt Water Desalination, Elsevier Science B.V., 2002.
[13] W.B. Glover, Selecting evaporators for process applications, Chem. Eng. Prog. 100 (December 2004) 26–33.
[14] B. Rahimi, K. Regenauer-Lieb, H.T. Chua, E. Boom, S. Nicoli, S. Rosenberg, A novel low grade heat driven process to re-concentrate process liquor in alumina refineries, Hydrometallurgy (2015) 327–336.
[15] B. Rahimi, K. Regenauer-Lieb, H.T. Chua, E. Boom, S. Nicoli, S. Rosenberg, A novel flash boosted evaporation process for alumina refineries, Appl. Therm. Eng. 94 (2016) 375–384.
[16] B. Rahimi, A. Christ, K. Regenauer-Lieb, H.T. Chua, A novel process for low grade heat driven desalination, Desalination 351 (October 2014) 202–212.
[17] X. Wang, A. Christ, K. Regenauer-Lieb, K. Hooman, H.T. Chua, Low grade heat driven multi-effect distillation technology, Int. J. Heat Mass Transf. 54 (25–26) (December 2011) 5497–5503.
[18] B. Rahimi, K. Regenauer-Lieb, H.T. Chua, E. Boom, S. Nicoli, S. Rosenberg, A novel low grade heat driven process to re-concentrate process liquor in alumina refineries, in: 10th Int. Alumina Quality Workshop (AQW) Conference, Perth, Australia. April 19th–23rd, 2015, 2015, pp. 327–336.
[19] Spirax Sarco Co, The Steam and Condensate Loop Book, Spirax Sarco Co, 2007.
[20] Solvay Chemicals Co, Liquid Caustic Soda Characteristics, 2014 (Online). Available: http://www.solvaychemicals.com/EN/products/causticsoda/Liquidcausticsoda.aspx.
[21] W.L. McCabe, The enthalpy - concentration chart – a useful device for chemical engineering calculations, Trans. AIChE 31 (1935) 129–169.
[22] J.W. Bertetti, W.L. McCabe, Sodium hydroxide solutions, Ind. Eng. Chem. 28 (2) (1936) 247–248.

[23] E.W. Washburn, C.J. West, C. Hull, National Academy of Sciences (U.S.), International Council of Scientific Unions., and National Research Council (U.S.), in: International Critical Tables of Numerical Data, Physics, Chemistry and Technology, vol. III, McGraw-Hill, New York, 1928.
[24] R.H. Perry, D.W. Green, J.O. Maloney, Perry's Chemical Engineers' Handbook, seventh ed., McGraw-Hill, 1999.
[25] B. Rahimi, J. May, A. Christ, K. Regenauer-Lieb, H.T. Chua, Thermo-economic analysis of two novel low grade sensible heat driven desalination processes, Desalination 365 (2015) 316–328.
[26] B. Rahimi, K. Regenauer-Lieb, H.T. Chua, A novel desalination design to better utilise low grade sensible waste heat resources, in: IDA World Congress 2015 on Desalination and Water Reuse, San Diego, US. August 30–September 4, 2015, 2015.
[27] A. Seifert, K. Genthner, A model for stagewise calculation of non-condensable gases in multi-stage evaporators, Desalination 81 (1991) 333–347.
[28] C. Sommariva, H. Hogg, K. Callister, Cost reduction and design lifetime increase in thermal desalination plants: thermodynamic and corrosion resistance combined analysis for heat exchange tubes material selection, Desalination 158 (May 2003) 17–21.
[29] C. Sommariva, H. Hogg, K. Callister, Maximum economic design life for desalination plant: the role of auxiliary equipment materials selection and specification in plant reliability, Desalination 153 (2002) 199–205.
[30] B. Rahimi, H.T. Chua, A. Christ, System and Method for Desalination, World Intellectual Property Organization, 2015. WO 2015/154142 A1.
[31] A. Christ, K. Regenauer-Lieb, H.T. Chua, Development of an advanced low-grade heat driven multi effect distillation technology, in: IDA World Congress on Desalination and Water Reuse, 2013.
[32] H.T. Chua, K. Regenauer-Lieb, X. Wang, A Desalination Plant, World Intellectual Property Organization, 2012. WO 2012/003525 A1.
[33] A. Christ, K. Regenauer-Lieb, H.T. Chua, Boosted Multi-Effect distillation for sensible low-grade heat sources: a comparison with feed pre-heating multi-effect distillation, Desalination 366 (2015) 32–46.
[34] A. Christ, K. Regenauer-Lieb, H.T. Chua, Thermodynamic optimisation of multi-effect-distillation driven by sensible heat sources, Desalination 336 (2014) 160–167.
[35] L.S. Lasdon, A.D. Waren, A. Jain, M. Ratner, J. Rice, Design and testing of a generalized reduced gradient code for nonlinear programming, ACM Trans. Math. Softw. 4 (1) (1978) 34–50.
[36] Alfa Laval, Single Effect Freshwater Generator, Model JWP-16/26-c Series, in Alfa Laval Marine & Diesel Product Catalogue, Alfa Laval Corporate AB, 2003.
[37] Kadant Thermocompressor Sizing Catalogue, Kadant Johnson Inc., MI, USA, 2011. http://www.kadant.com/.
[38] NIST, NIST Reference Fluid Thermodynamic and Transport Properties Database (REFPROP): Version 9.1, 2013.
[39] J.W. Bertetti, W.L. McCabe, Specific heats of sodium hydroxide solutions, Ind. Eng. Chem. 28 (3) (1936) 375–378.
[40] H.R. Wilson, W.L. McCabe, Specific heats and heats of dilution of concentrated sodium hydroxide solutions, Ind. Eng. Chem. 34 (5) (1942) 558–566.
[41] IDA Desalting Plants Inventory, 2011 (Online). Available: http://www.desaldata.com.

APPENDIX A

Seawater Enthalpy

Latent heat of evaporation [1]:

$$h_{fg(T_{sv})} = 2499.5698 - 2.204864 \cdot T_{sv} - 1.596 \times 10^{-3} \cdot T_{sv}^{2} \tag{A.1}$$

Seawater enthalpy [2]:

$$h_{f,sw} = h_{f,w} - X \cdot (a_1 + a_2 \cdot X + a_3 \cdot X^2 + a_4 \cdot X^3 + a_5 \cdot T_B + a_6 \cdot T_B^2 + a_7 \cdot T_B^3 + a_8 \cdot X \cdot T_B + a_9 \cdot X^2 \cdot T_B + a_{10} \cdot X \cdot T_B^2) \tag{A.2}$$

which is valid for $10 \leq T_B \leq 120°C$ and $0 \leq X \leq 0.12$ kg/kg range with an accuracy of $\pm 0.5\%$. $h_{f,w}$ is the saturated enthalpy of pure water:

$$h_{f,w} = 141.355 + 4202.07 \cdot T_B - 0.535 \cdot T_B^{2} + 0.004 \cdot T_B^{3} \tag{A.3}$$

which is valid for $5 \leq T_B \leq 200°C$ and an accuracy of $\pm 0.02\%$. The constants are as follows:

$$a_1 = -2.348 \times 10^4;\ a_2 = 3.152 \times 10^5;\ a_3 = 2.803 \times 10^6;\ a_4 = -1.446 \times 10^7;$$
$$a_5 = 7.826 \times 10^3;\ a_6 = -4.417 \times 10^1;\ a_7 = 2.139 \times 10^{-1};\ a_8 = -1.991 \times 10^4;$$
$$a_9 = 2.778 \times 10^4;\ a_{10} = 9.728 \times 10^1$$

References

[1] H.T. El-Dessouky, I. Alatiqi, S. Bingulac, H.M. Ettouney, Steady-state analysis of the multi effect evaporation desalination process, Chem. Eng. Technol. 21 (1998) 437–451.

[2] M.H. Sharqawy, J.H. Lienhard V, S.M. Zubair, Thermophysical properties of seawater: a review of existing correlations and data, Desalin. Water Treat. 16 (10) (2010) 354–380.

Boiling Point Elevation and Nonequilibrium Allowance

Boiling point elevation (*BPE*) for seawater is calculated by Ref. [2] of, Appendix A

$$BPE = \lambda \cdot X^2 + \sigma \cdot X \tag{B.1}$$

$$\lambda = -4.584 \times 10^{-4} \cdot {T_B}^2 + 2.823 \times 10^{-1} \cdot T_B + 17.95 \tag{B.2}$$

$$\sigma = 1.536 \times 10^{-4} \cdot {T_B}^2 + 5.267 \times 10^{-2} \cdot T_B + 6.56 \tag{B.3}$$

The validity of Eq. (B.1) is for $0 \leq T_B \leq 200°\text{C}$; $0 \leq X \leq 0.12$ kg/kg range with an accuracy of $\pm 0.018\text{K}$.

BPE of caustic soda (NaOH), which is mentioned in Chapter 9, can be found in Fig. B.1 [1].

Nonequilibrium allowance (*NEA*) for flashing chambers is calculated by Ref. [2]

$$NEA_i = \frac{50.7 \cdot \Delta P'^{0.71} \cdot SWL^{0.07} \cdot {H_i}^{1.1}}{{T_{FC,i,in}}^{2.0} \cdot {L_i}^{1.01}} \tag{B.4}$$

where $\Delta P'$ is the vapor pressure difference in the chamber and is calculated by

$$\Delta P' = P_{sat\left\langle T_{FC,i,in} \right\rangle} - P_i \tag{B.5}$$

SWL is the specific weir load in (kg/s)/m and is expressed as [3]

$$SWL = 105.6 + 18.06 \cdot D_{FC} \tag{B.6}$$

where [4]

$$D_{FC} = \sum_{i=1}^{j} \dot{m}_{v,FC,i} \tag{B.7}$$

L_i is the chamber length [5]:

$$L_i = \frac{\dot{m}_{V,FC,i}}{V_v \cdot W_i \cdot \rho_{v,i}}$$

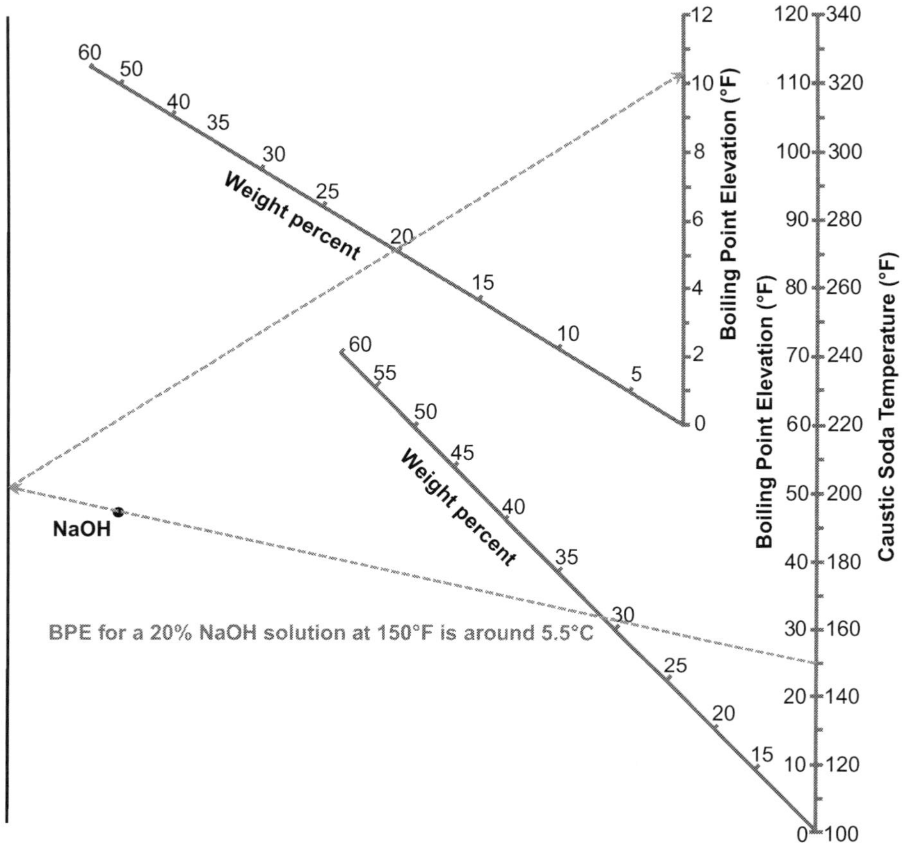

Figure B.1
Boiling point elevation of caustic soda [1].

where V_v is the maximum allowable vapor released velocity, which is limited to 6 m/s; ρ_v is the vapor density at $T_{sat\langle P_i \rangle}$; and W_i is the chamber width, which is calculated by

$$W_i = \frac{\dot{m}_{FC,i,in}}{SWL} \tag{B.8}$$

H_i is the average brine level (m), which is around 0.1−0.2 m higher than the sluice gate height (HG_i) [4,5]:

$$H_i \approx 0.1 + HG_i \tag{B.9}$$

and [4]

$$HG_i = \frac{\dot{m}_{FC,i,in}}{C_d \cdot W_i} \cdot \left(2 \cdot \rho_{w,i} \cdot \Delta P_{v,i}\right)^{-0.5} \tag{B.10}$$

where $\rho_{w,i}$ is the density of pure water at $T_{sat\langle P_i \rangle}$, $\Delta P_{v,i}$ is the vapor pressure difference, and C_d is the weir discharge coefficient, which is considered to be 0.6 for free flow and 0.7 for submerged flow [6].

References

[1] R.H. Perry, D.W. Green, J.O. Maloney, Perry's Chemical Engineers' Handbook, seventh ed., McGraw-Hill, 1999.

[2] R. Rautenbachr, S. Schafer, S. Schleiden, Improved equations for the calculation of non-equilibrium temperature loss in MSF, Desalination 108 (1996) 325–333.

[3] H. El-Dessouky, S. Bingulac, A fixed point iterative algorithm for solving equations modeling the multi-stage flash desalination process, Comput. Methods Appl. Mech. Eng. 141 (1–2) (February 1997) 95–115.

[4] H. El-Dessouky, S. Bingulac, Solving equations simulating the steady-state behavior of the multi-stage flash desalination process, Desalination 107 (2) (October 1996) 171–193.

[5] H. El-Dessouky, H.I. Shaban, H. Al-Ramadan, Steady-state analysis of multi-stage flash desalination process, Desalination 103 (1995) 271–287.

[6] M.A. Darwish, M.M. El-Refaee, M. Abdel-Jawad, Developments in the multi-stage flash desalting system, Desalination 100 (1995) 35–64.

APPENDIX C

Pressure Drop Across Plate Heat Exchangers

The pressure drop across a plate heat exchanger includes three components:

1. Pressure drop associated with the distribution inlet and outlet manifolds and ports, which should be kept as low as possible
2. Pressure drop across the plates' passages (the core)
3. Static head due to the elevation change

The pressure drop on each side of a plate heat exchanger for water–water application can be estimated by Ref. [1]

$$\Delta P_{phe} = \frac{1.5G^2N}{2\rho_{inn}} + \frac{4\zeta H_{plate}G^2}{2\rho_m d_e} \pm \rho_m g H_{plate} \tag{C.1}$$

where N is the number of passes on each side, $G = 4\dot{m}/\left(\pi {d_e}^2\right)$ is the fluid mass velocity in the port; d_e is the equivalent diameter of flow passages, which can be consider as twice the plate spacing; ζ is Fanning's friction factor, which is a function of Re number and is usually determined by empirical correlations; ρ_i is the fluid density at the local inlet bulk temperature; and $\rho_m = (\rho_{inn} + \rho_o)/2$ is the mean density. In Eq. (C.1) the unit of ΔP_{phe} is in Pa.

To simplify the calculation for water-to-water application, we can consider 0.20–0.98 bar pressure drop per number of transfer units (NTU) [2], which is a very rough but conservative consideration. In this book for the liquid–liquid plate heat exchangers a 0.3 bar per *NTU* has been considered in the calculations [3]. For evaporators and condensers a 0.3 bar and a 0.4 bar per *NTU* have been considered, respectively [2].

NTU of each side of the heat exchangers can be calculated by

$$NTU_{hot} = \frac{\left|T_{hot,in} - T_{hot,out}\right|}{LMTD} \tag{C.2}$$

$$NTU_{cold} = \frac{\left|T_{cold,in} - T_{cold,out}\right|}{LMTD} \tag{C.3}$$

where

$$LMTD = \frac{\left(T_{hot,in} - T_{cold,out}\right) - \left(T_{hot,out} - T_{cold,in}\right)}{\ln\left(\dfrac{T_{hot,in} - T_{cold,out}}{T_{hot,out} - T_{cold,in}}\right)} \tag{C.4}$$

Therefore, the pressure drop (in bar) for each hot and cold side of a water–water plate heat exchanger with a conservative approximation can be estimated by

$$\Delta P_{hot} \approx 0.3 \cdot NTU_{hot} \tag{C.5}$$

$$\Delta P_{cold} \approx 0.3 \cdot NTU_{cold} \tag{C.6}$$

References

[1] R.K. Shah, D.P. Sekulic, Fundamentals of Heat Exchanger Design, John Wiley & Sons, Inc., 2003.
[2] O. Lekang, Aquaculture Engineering, Blackwell Publishing, 2007.
[3] T. Kuppan, Heat Exchanger Design Handbook, Marcel Dekker Inc., 2000.

APPENDIX D

Overall Heat Transfer Coefficient in Condenser and Falling Film Evaporators

The calculation of the overall heat transfer coefficients across the first and the second effect and the condenser of the flash-boosted thermal vapor compression multi-effect evaporation (FB-TVC-MEE) process are explained as follows.

The overall heat transfer coefficient across the condenser can be calculated from Eq. (D.1) as a function of the saturated temperature of the inlet vapor [1], which is applicable to both TVC-MEE and FB-TVC-MEE processes:

$$U_{cond.} = 1.7194 + 3.2063 \times 10^{-2} \times T_{sat,cond.} - 1.5971 \times 10^{-5} \times \left(T_{sat,cond.}\right)^2 + 1.9918 \times 10^{-7} \times \left(T_{sat,cond.}\right)^3 \quad \text{(D.1)}$$

The general equation for the overall heat transfer coefficient of the falling film evaporators (first and second effects) is written as

$$\frac{1}{U} = \left(\frac{1}{\alpha_{inn}} + R_{F,inn}\right)\cdot\frac{r_o}{r_{inn}} + \frac{r_o}{k_{tube}}\cdot\ln\left(\frac{r_o}{r_{inn}}\right) + \frac{1}{\alpha_o} + R_{F,o} \quad \text{(D.2)}$$

In our case the overall heat transfer coefficient with a reasonable approximation can be considered as a function of the film coefficients (α_i and α_o) since the conduction resistance of the wall and fouling resistance ($R_{F,i}$ and $R_{F,o}$) are considerably smaller. r_o/r_i has been considered to be 1.2 in our calculations, based on typical diameter range for schedule 40 tube evaporators.

According to Ref. [2], α_i is calculated based on Eq. (D.3).

$$\alpha_i = \alpha^+ \cdot \left(\frac{\mu_L{}^2}{\rho_L{}^2 \cdot k_L{}^3 \cdot g}\right)^{-\frac{1}{3}} \quad \text{(D.3)}$$

where

$$\alpha^+ = 1.6636 \cdot Re_L{}^{-0.2648} \cdot Pr_L{}^{-0.1592};\quad 15 < Re_L < 3000;\quad 2.5 < Pr_L < 200 \quad \text{(D.4)}$$

Table D.1: Heat transfer coefficients in kW/(m^2·K) for TVC-MEE and FB-TVC-MEE processes (see, Chapter 9).

	TVC-MEE			FB-TVC-MEE		
	1st Effect	2nd Effect	Condenser	1st Effect	2nd Effect	Condenser
α_i	3.8	3.3	–	3.7	3.2	–
α_o	4.7	4.2	–	4.7	4.2	–
U	1.9	1.6	3.2	1.9	1.6	3.2

FB-TVC-MEE, flash-boosted thermal vapor compression multi-effect evaporation; *TVC-MEE*, thermal vapor compression multi-effect evaporation.

To simplify the calculation for water-to-water application, we can consider 0.20–0.98 bar pressure drop per number of transfer units (NTU).

The relevant heat transfer coefficient for the steam condensation on the outer tube can be calculated from Eq. (D.5) or (D.6) [3,4].

$$\alpha_o = \frac{Re_L \cdot k_L}{1.08 \cdot {Re_L}^{1.22} - 5.2} \cdot \left(\frac{g}{{v_L}^2}\right)^{1/3}; \quad 30 < Re_L < 1800 \tag{D.5}$$

$$\alpha_o = \frac{Re_L \cdot k_L}{8750 + 58 \cdot {Pr_L}^{-0.5} \cdot \left({Re_L}^{0.75} - 253\right)} \cdot \left(\frac{g}{{v_L}^2}\right)^{1/3}; \quad Re_L > 1800 \tag{D.6}$$

The calculated heat transfer coefficients are shown in Table D.1.

References

[1] I.S. Al-Mutaz, I. Wazeer, Development of a steady-state mathematical model for MEE-TVC desalination plants, Desalination 351 (2014) 9–18.

[2] J.S. Prost, M.T. González, M.J. Urbicain, Determination and correlation of heat transfer coefficients in a falling film evaporator, J. Food Eng. 73 (2006) 320–326.

[3] A. Yunus, Çengel, Heat Transfer: A Practical Approach, McGraw-Hill, 2003.

[4] T.L. Bergman, A.S. Lavine, F.P. Incropera, D.P. DeWitt, Fundamentals of Heat and Mass Transfer, seventh ed., John Wiley & Sons, Inc., 2011.

APPENDIX E

Plate Heat Exchanger Cost Estimation

The capital cost of a heat exchanger is a function of area and the material. There are many chemical engineering references that include the cost function for process equipment such as heat exchangers, evaporators, vessels, cooling towers, pumps, and so on. The function that is used in this book to estimate the plate and frame heat exchanger capital cost is based on Ref. [1].

This function is used for the plate and frame heat exchanger with a design pressure not more than 10 bar and a design temperature up to 160°C.

For areas less than 18.6 m^2:

$$CC_{hex} = 1281 \cdot IF \cdot A^{0.4887} \text{ for 316 stainless steel} \tag{E.1}$$

$$CC_{hex} = 1839 \cdot IF \cdot A^{0.4631} \text{ for titanium (grade 1)} \tag{E.2}$$

For areas more than 18.6 m^2:

$$CC_{hex} = 702 \cdot IF \cdot A^{0.6907} \text{ for 316 stainless steel} \tag{E.3}$$

$$CC_{hex} = 781 \cdot IF \cdot A^{0.7514} \text{ for titanium (grade 1)} \tag{E.4}$$

In the previous equations, A is the area in m^2; IF is the installation factor that can be considered to be between 1.5 and 2.0 (we have used 2.0 in this book), depending on the size of unit; and CC_{hex} is the capital cost of plate and frame heat exchanger in \$US.

Reference

[1] C. Haslego, G. Polley, Compact heat exchangers – Part 1: Designing plate-and-frame heat exchangers, Chem. Eng. Prog. 98 (9) (2002) 32–37.

Plate Heat Exchanger Overall Heat Transfer Coefficient

To find the overall heat transfer coefficient in plate heat exchanger, use Eq. (F.1):

$$\frac{1}{U} = \frac{1}{\alpha_{hot}} + R_{F,hot} + \frac{\delta}{k_{plate}} + \frac{1}{\alpha_{cold}} + R_{F,cold} \tag{F.1}$$

where U is the overall heat transfer coefficient, α_{hot} and α_{cold} are the local heat transfer coefficients for the hot and cold sides, δ is the plate thickness, k_{plate} is the conduction resistance of the plate, and $R_{F,hot}$ and $R_{F,cold}$ are the relevant fouling factors of each side, which for simplicity can be ignored.

α_{hot} and α_{cold} can be calculated based on Figs. F.1–F.3 of [1], which show the local heat transfer coefficients as a function of pressure drop and the average viscosity. k_{plate} for titanium is considered to be around 16 W/(m·K). In our case a 0.4 mm plate thickness has been considered.

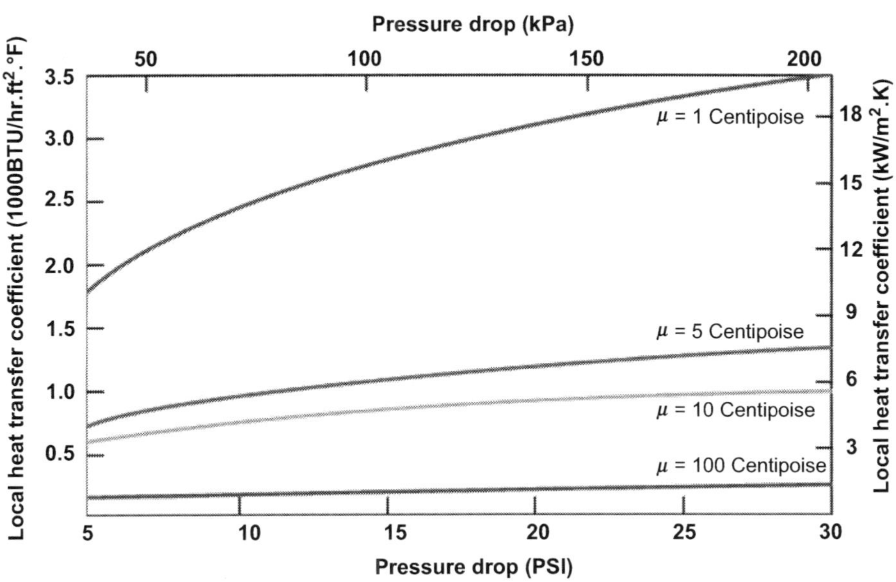

Figure F.1
Local heat transfer coefficient for water-based fluids, 0.25 < NTU < 2.0 [1].

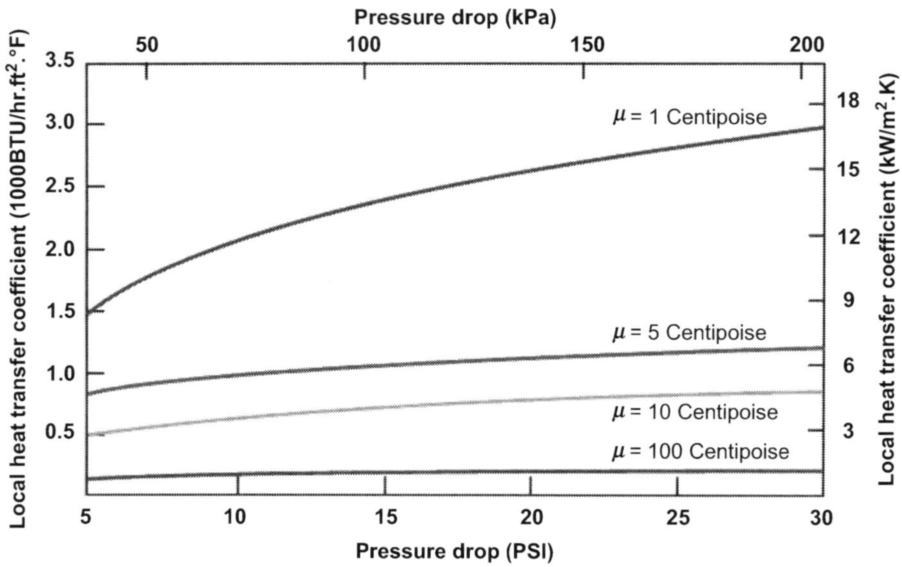

Figure F.2

Local heat transfer coefficient for water-based fluids, $2.0 < NTU < 4.0$ (for NTU between 3.5 and 4.0, local heat transfer coefficient should be reduced by 15% for liquids with viscosities less than 2.0 centipoise) [1].

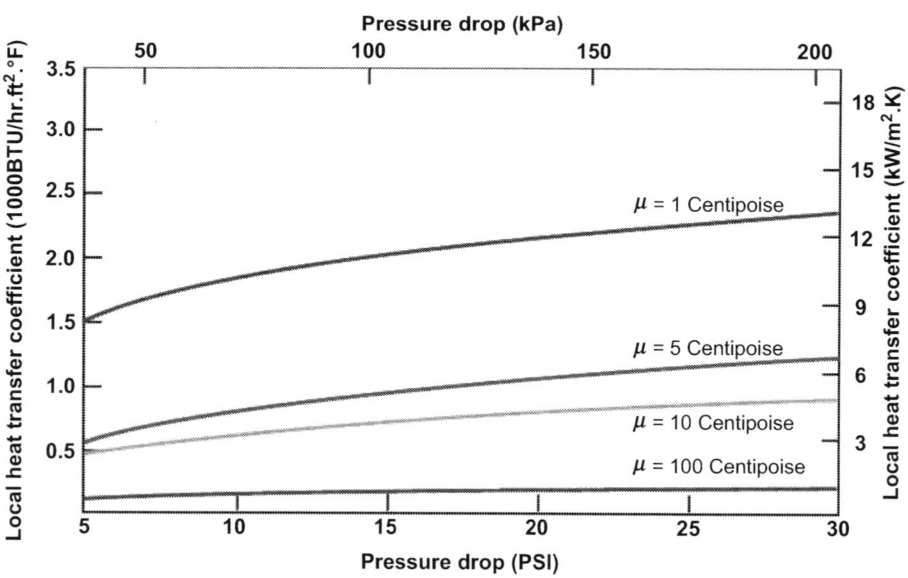

Figure F.3

Local heat transfer coefficient for water-based fluids, $4.0 < NTU < 5.0$ [1].

Reference

[1] C. Haslego, G. Polley, Compact heat exchangers – Part 1: Designing plate-and-frame heat exchangers, Chem. Eng. Prog. 98 (9) (2002) 32–37.

APPENDIX G

Excel Spreadsheet

G.1 Introduction

This spreadsheet is a student version platform and includes both the technical and economic analyses for conventional multi-effect distillation (MED), boosted MED (B-MED), and flash-boosted MED (FB-MED) processes for seawater application and 65°C heat source inlet temperature scenario. The results provided in this book stem from an original version of this solver (not the student version), which used the REFPROP package (REFerence fluid PROPerties) as developed by the National Institute of Standards and Technology (NIST) [1]. The student version spreadsheet uses the Water97_13 add-in instead of REFPROP.

This spreadsheet uses the Excel solver [2] and includes seven worksheets (Table G.1).

G.2 Thermoeconomic Simulation Worksheets

As mentioned earlier, the first three worksheets are related to the thermoeconomic analysis of all the previously mentioned processes. These worksheets are identical in principle and include the sections as listed next. Each cell has been linked to other cells based on a

Table G.1: Worksheet descriptions.

Worksheet title	Description
Conventional MED	It includes a thermoeconomic simulation for a conventional MED process.
B-MED	It includes a thermoeconomic simulation for a conventional B-MED process.
FB-MED	It includes a thermoeconomic simulation for a conventional FB-MED process.
MED cash flow analysis	It includes a cash flow analysis for the conventional MED scenario.
B-MED cash flow analysis	It includes a cash flow analysis for the B-MED scenario.
FB-MED cash flow analysis	It includes a cash flow analysis for the FB-MED scenario.
Choosing the best alternative	It includes an NPV-analysis graph, which helps the user choose the best alternative.

B-MED, boosted multi-effect distillation; *FB-MED*, flash-boosted multi-effect distillation; *MED*, multi-effect distillation; *NPV*, net present value.

relevant equation, which is explained in detail in Chapters 4 to 7. These worksheets include eight sections as listed here.

1. INITIAL INPUTS
   In this section, the user can enter the inputs. These values can be changed to simulate different scenarios.
2. INITIAL GUESS
   In this section, the user can enter guessed parameters. These parameters are identified based on the process specifications. In order to simulate the FB-MED process, the number of initial guesses is more than the other processes. Remember that a good guess can lead the solver to solve quickly, so the user should have some preliminary engineering understanding about the process to offer a good guess for the relevant parameters.
3. ASSUMPTIONS
   This section includes the secondary inputs. The user can change them (in a logical and practical way) to simulate different possibilities. For example the feed-to-vapor ratio can be changed from 2.857 (which stands for 35% evaporation) to 2.5 (40% evaporation), or the minimum temperature difference across each primary MED effect can be considered to be 4°C instead of 2.5°C, and so on.
4. OUTPUTS
   This section includes more important calculated parameters such as the total production rate, specific pumping power, waste heat performance ratio, and so on.
5. CAPEX and OPEX
   This section includes capital and operating cost estimation, which is detailed in Chapter 7.
6. PUMPING POWER EVALUATION
   This section includes pumping power analyses. By changing the pressure drops or efficiencies, the total pumping power can be changed (see Chapter 5).
7. MASS and ENERGY BALANCE
   This section includes mass and energy balance. The cells of this section should not be changed because they are linked together and with other cells. After solving the simulation, the user should check the balances to be sure about the results. Sometimes the solver solves without any error and reaches a solution, but the mass and energy in some sections of the process are not balanced. In this situation the user should double-check the simulation and solve the problem.
8. THE MAIN PROCESS
   This section includes the effects and all heat transfer devices such as flashing chambers, heat exchangers, condenser, and deaerator. Each part includes a primary colored box that shows the essential calculated parameters such as temperatures and flow rates. Under this box the other parameters are shown such as UA values, enthalpies, pressures and pressure drops, salinities, boiling point elevation (BPE), and so on. These parameters should not be changed because all of them are linked together and with the input data. These parameters are shown to provide a better understanding of the process for the user. The boundary conditions of each relevant section such as first effect, condenser, booster (for B-MED process), flashing chambers and the heat exchanger (for FB-MED process) are located in the last box. The solver uses these boundary conditions based on the explanations in Section 4.2.6).

G.3 Cash Flow Analysis Worksheets

These worksheets include a very detailed cash flow analysis for the entire life of a plant. They are identical for all the above mentioned processes.

1. INPUTS
   This section includes the required parameters for a cash flow analysis (see Chapter 7). The INPUTS section of the MED Cash Flow Analysis worksheet can be changed by the user and the INPUTS section of the other processes are linked to these parameters and should not be changed or modified.
2. PLANT SPEC.
   This section includes some specifications of the plant, which are calculated automatically and should not be changed.
3. OUTPUTS
   This section shows the important outputs such as net present value (NPV), payback period, and so on.
4. RATIOS
   This section compares the capital and operating cost estimation, and the production rate of the processes. The conventional MED process is the benchmark.
5. Cash Flow Table
   This table shows a detailed cash flow during the project lifetime in each year. This table helps the user to understand the cash flow analysis in a better way.

G.4 Choosing the Best Alternative Worksheet

This worksheet uses NPV, internal rate of return (IRR), and incremental rate of return (ΔIRR) methods to help make a decision among the processes. It also plots an NPV versus interest rate graph. See Chapter 7 for more details.

G.5 Relevant Add-Ins

This simulation needs some add-ins that need to be run, as follows:

1. **Seawater16_v1**: This add-in includes the properties of seawater such as enthalpy, density, and BPE based on Ref. [2] in Appendix A. The relevant functions of this add-in are as follows. seawaterEnthalpy (X,T), seawaterDensity(X,T), and seawaterBPE(X,T), where X is the seawater salinity in (kg/kg) and T is the seawater temperature in (°C).
2. **Vacuum_Pumping_Power**: This add-in calculates the pumping power consumption of the liquid-ring vacuum pump in kW. The considered data are based on Robuschi Vacuum pump products [3]. The relevant function of this add-in is pumpNa$(P_{vac}, \dot{V}_{NCG})$, where P_{vac} is the absolute suction pressure of the pump in (mbar) and $\dot{V}_{NCG}$ is the saturated air capacity at 20°C in (m^3/h). This function is valid for $33 < P_{vac} < 800$ mbar and $19 < \dot{V}_{NCG} < 3500$ m^3/h (depending on the pressure).

Table G.2: Meaning of important colors.

Colour	Description
Orange	Cells with this colour means inputs that the user can enter, change or modify based on logical and theoretical or practical engineering values.
Green	Green cells are for initial guesses.
Blue-Gray	In case of adding or removing one or more effects or flashing chambers, the equation of these cells should be updated.

3. **Water97_13**: This add-in calculates all the water and steam properties; for more details about its functions, refer to http://www.cheresources.com/iapwsif97.shtml. However, in the original simulator REFPROP NIST [1] has been used instead of this freeware.

G.6 Important Colors

For simplicity of finding the appropriate cells, some colors have been used as shown in Table G.2.

G.7 Adding One More Effect

The user has the ability to add one or more effects or flashing chambers and prepare for different scenarios. This procedure is simple but it should be done carefully. The user should check the links between the new effects or flashing chambers' relevant cells.

For example, to add one effect to the conventional MED worksheet and increase the number of effects to 6, the following steps should be considered.

1. Select the condenser section (a box from Cell X2 to Z50). Cut and paste to Cell AA2 to AC50.
2. Select the last effect (effect no. 4 in this case): a box from Cell U2 to W37. Copy and paste to Cell X2 to Z37.
3. Change the name of this new effect to Effect 5.
4. Change the value of Cell D7 to 5.
5. Now the condenser's cells should be linked to the new effect. For this purpose:
   a. Select Cell AA5 and link it to Cell Z9 instead of Cell W9.
   b. Select Cell AA7 and link it to Cell Z7.
   c. Select Cell AA11 and link it to Cell Z13.
   d. Select Cell AC27 and update it with Cell AA5 instead of X5 and Cell Z37 instead of W37.

6. Update all the blue-gray highlighted cells such as H3, H8, H9, Z17, and some others in the Mass and Energy Balance section.
7. Update the Mass and Energy Balance section by adding energy balance of the added effect.
8. Run the Solver.

The same method will be applied for the B-MED and FB-MED worksheets. By adding or removing one effect to the primary MED section, the relevant cell for the condenser should be linked to the new effect.

In order to change the injection point of the booster, the relevant vapor mass flow rate of the injected effect should be updated. Moreover, the relevant injected vapor temperature cells in the boundary condition section of the booster module should be modified and linked to the relevant injected effect's pressure.

In order to add a flashing chamber in the FB-MED worksheet, the same method should be applied. It should be noted that each flashing chamber is linked to a relevant MED effect, so that all these connections should be updated or prepared (for each new effect and flashing chamber). Moreover, its relevant boundary conditions should be added to the Solver.

References

[1] NIST, NIST Reference Fluid Thermodynamic and Transport Properties Database (REFPROP): Version 9.1, 2013.

[2] Solver. [Online] Available: http://www.excel-easy.com/data-analysis/solver.html.

[3] S.R.L. Gardner Denver, Robuschi Liquid Vacuum Pumps Catalogue, 2016 [Online] Available: http://www.gardnerdenver.com/robuschi/downloads/vacuum-pumps/.

Nomenclature

A	Evaporator area (m^2)
AEC	Annual electrical cost (US$/yr)
ATAX Income	Annual after-tax income (US$/year)
a	Constant
B-MED	Boosted multi-effect distillation
BLP	Percentage of desalination plant capital cost that is covered by loan
BPE	Boiling point elevation (°C)
CA	Cost of asset (US$)
CC	Capital cost (US$)
ChC	Annual chemical cost (US$/year)
CMP	Compression ratio
CRF	Capital recovery factor
CUP	Chemical unit price (US$/$m^3$)
$\mathbf{C_d}$	Weir discharge coefficient
$\mathbf{d_e}$	Equivalent diameter of flow passage (m)
$\mathbf{D_{FC}}$	Total process condensate production rate (kg/s)
$\mathbf{D_t}$	Total process condensate production rate (m^3/day)
DR	Drain ratio
E	Energy (kJ)
ENT	Entrainment ratio
ES	Escalation factor
EUP	Electricity unit price (US$/kWh)
EXP	Expansion ratio
FB-MED	Flash-boosted multi-effect distillation
FB-MEE	Flash-boosted multi-effect evaporation
FB-TVC-MEE	Flash-boosted thermal vapor compression multi-effect evaporation
f	Plant availability
g	Acceleration of gravity (m/s^2)
G	Fluid mass velocity ($kg/(m^2 \cdot s)$)
GOR	Gain output ratio
GRG	Generalized reduced gradient
$\Delta\dot{\mathbf{H}}$	Heat transfer rate (MW)
$\Delta\mathbf{h_{available}}$	Maximum exploitable energy of the heat source (kJ/kg)
$\Delta\mathbf{h_{ref}}$	Specific reference enthalpy of the distillate (kJ/kg)
H	Brine level (m)
HG	Sluice gate height (m)
$\mathbf{H_{plate}}$	Plate height (m)
h	Enthalpy (kJ/kg)
$\mathbf{h_f}$	Enthalpy of saturated liquid (kJ/kg)
$\mathbf{h_{fg}}$	Enthalpy of evaporation/condensation (latent heat) (kJ/kg)

h_g	Enthalpy of saturated or superheated steam (kJ/kg)
h'_g	Enthalpy of saturated or superheated steam (kJ/kg)
ΔIRR	Incremental rate of return (%)
IF	Installation factor
Income	Annual income of the plant (US$/year)
IPMT	Annual bank interest payment (US$/year)
IRR	Internal rate of return (%)
ITAX	Income tax (US$/year)
i	Annual interest rate
k	Thermal conductivity (W/(m·K))
L	Flash stage length (m)
LA	Total loan amount (US$)
LC	Annual labor cost (US$/year)
LMTD	Logarithmic mean temperature difference (°C)
MARR	Minimum acceptable rate of return
MED	Multi-effect distillation
MEE	Multi-effect evaporation
MSF	Multi-stage flash
MSIC	Annual maintenance, spares, and insurance costs (US$/year)
$\dot{m}$	Mass flow rate (kg/s)
$\dot{m}_{D,total}$	Total process condensate (freshwater) mass flow rate (kg/s)
$\dot{m}_m$	Motive steam flow rate (kg/s)
$\dot{m}_p$	Total produced process condensate excluding the injected vapor from the waste stream (kg/s)
N	Number of passes on each side of a plate heat exchanger
NCG	Noncondensable gasses
NEA	Nonequilibrium allowance (°C)
NEA_{10}	NEA for a 10-ft flashing stage (°C)
NPV	Net present value (US$)
NTU	Number of transfer units
n	Desalination plant operating life (year)
OPEX	Operating expenditure (US$/year)
ΔP	Pressure difference (kPa)
$\Delta P_{v,i}$	Vapor pressure difference between the adjacent stages (Pa)
P	Pressure (bar) or (kPa)
Pr	Prandtl number
P-MED	Preheated multi-effect distillation
PMT	Loan repayment per year (US$/year)
PPMT	Annual bank principal payment (US$/year)
PR	Performance ratio
PR_{WH}	Waste heat performance ratio
p	Specific pumping power (kWh/m^3)
$\dot{Q}$	Heat transfer rate (kW)
R	Feed-to-vapor ratio
RF	Recovery factor
R^2	R-squared value

$\mathbf{R_F}$	Fouling resistance ($m^2 \cdot K/kW$)
Re	Reynolds number
r	Evaporator tube radius (m)
S	Supplementary fee on loan as a percentage of the loan amount
SCC	Specific capital cost (US$/($m^3$/day))
SL	Simple straight line depreciation (US$/year)
SV	Salvage value (US$)
SWL	Specific Weir load (kg/(s·m))
Δ**T**	Temperature difference (°C)
T	Temperature (°C)
TBT	Top brine temperature (°C)
TDS	Total dissolved solids
TCC	Total capital cost (US$)
TR	Tax rate
TVC-MEE	Thermal vapor compression multi-effect evaporation
U	Overall heat transfer coefficient (kW/($m^2 \cdot K$))
UA	UA Value (kW/K)
UPC	Unit product cost (US$/$m^3$)
$\mathbf{V_v}$	Maximum allowable vapor released velocity (m/s)
$\dot{\mathbf{V}}$	Volumetric flow rate (m^3/s)
$\dot{\mathbf{V}}_{\mathbf{D,total}}$	Total production rate of the plant (m^3/h)
W	Flash stage width (m)
WMP	The unit market price of water (US$/$m^3$)
X	Concentration (wt%); salinity (ppm)
x	Steam quality
α	Heat transfer coefficient (kW/($m^2 \cdot K$))
α^+	Dimensionless heat transfer coefficient
η	Overall pump efficiency
θ	Flag distinguishing FB-TVC-multi-effect evaporation from TVC-multi-effect evaporation
λ	Constant
δ	Plate thickness (m)
μ	Viscosity (Pa·s)
ρ	Density (kg/m^3)
σ	Constant
ν	Kinematic viscosity (m^2/s)
ζ	Fanning's friction factor
ϕ	Area coefficient
Υ	Overall area coefficient
Ψ	Cost function

Subscripts

0	Zeroth year
1	First effect; first year
2	Second effect; second year

3	Third effect; third year
4	Fourth effect; fourth year
B	High concentrate outlet or brine
B^*	Relevant boiling point
B^{**}	Relevant brine temperature
B-MED	Boosted multi-effect distillation process
BBP	Brine blow-down pump
BRP	Brine recirculation pump
bstr	Booster
C	Coolant (cooling water supply)
C^*	Relevant coolant temperature
c_hex	Cold side of the liquid–liquid heat exchanger
c_prh	Cold side of the preheater
calc.	Calculated
cold	Cold side
cond	Condenser
D	Distillate
$D_{Booster}$	Total generated vapor from the booster effect (m^3/day)
D_{FC}	Total flashed vapor production rate from flash chambers (m^3/day)
D_{FV}	Total flashed vapor production rate from flash vessels (m^3/day)
D_{inj}	Injected vapor flow rate (m^3/day)
D_t	Total process condensate production rate (m^3/day)
DEP	Distillate extraction pump
DRP	Drain pump
d	Discharge steam
e	Effect
F	Feed
F^*	Relevant feed temperature
FB-MED	Flash-boosted multi-effect distillation process
FB-MEE	Flash-boosted multi-effect evaporation process
FB-TVC-MEE	Flash-boosted thermal vapor compression multi-effect evaporation process
FC	Flashing chamber
HC	Highly concentrated process liquor
HS	Heat source (for effects), hot side (for the condenser)
HSP	Heat source medium pump
h_hex	Hot side of the liquid–liquid heat exchanger
h_prh	Hot side of the preheater
hex	Liquid–liquid heat exchanger
hot	Hot side
i	Flashing chamber number (flash-boosted multi-effect distillation process); preheater number (preheated multi-effect distillation process)
in	Inlet
inj	Injected
inn	Inner surface
j	Total number of flashing stages (flash-boosted multi-effect distillation process); total number of preheaters (preheated multi-effect distillation process)

k	Multi-effect distillation effect's number
k^*	Multi-effect distillation effect's number that has been injected by flashed vapor
L	Liquid
lm	Log mean
MED	Multi-effect distillation process
MEE	Multi-effect evaporation process
MK	Make-up
MKP	Make-up water pump
MSF	Multi-stage flash process
m	Motive steam; year number
n	Total number of multi-effect distillation effects; total plant's life in year
o	Outer surface
out	Outlet
P	Effect's pressure
P_1	First effect's pressure
P_2	Second effect's pressure
P_d	Discharge steam pressure
P_i	i^{th} flashing chamber's pressure
P_k	k^{th} effect's pressure
P_m	Motive steam pressure
P_s	Suction steam pressure
P_n	n^{th} effect's pressure
P-MED	Preheated multi-effect distillation process
p	Process condensate
phe	Plate heat exchanger
pinch	Pinch point
prh	Preheater
R	Recycled brine
SWP	Saline water pump
s	Suction steam
sat	Saturated steam
sv	Saturated vapor
sw	Seawater
s–l	Sensible–latent heat transfer
s–s	Sensible–sensible heat transfer
t	Total
T_d	Discharge temperature
v,V	Vapor
V'	Injected vapor (flashed vapor from flashing chambers)
vac	Vacuum
w	Pure water
z1	Zone 1 of the temperature–energy profile
z2	Zone 2 of the temperature–energy profile
z3	Zone 3 of the temperature–energy profile

Index

'*Note*: Page numbers followed by "f" indicate figures, "t" indicate tables.'

FSC
www.fsc.org
MIX
Papier | Fördert
gute Waldnutzung
FSC® C083411

Kolibri 360 GmbH
Ferdinand-Jühlke-Straße 7
99095 Erfurt, Deutschland
produktsicherheit@kolibri360.de